THE SHEEP STELL

The Sheep Stell

Memoirs of a Shepherd

JANET WHITE

Constable • London

CONSTABLE

Some names and details have been changed to protect the privacy of others.

First published in hardback in Great Britain in 1991 by The Sumach
Press, 29 Mount Pleasant, St Albans, Herts, AL3 4QY

First published in paperback in 2009 by Cantoc Books,
Durborough, Aisholt, Bridgwater, TA5 1AP

This revised edition published in 2018 by Constable

1 3 5 7 9 10 8 6 4 2

A CIP catalogue record for this book
is available from the British Library.

ISBN: 978-1-47212-860-7 (hardback)
ISBN: 978-1-47212-861-4 (trade paperback)

Typeset in Palatino by Hewer Text UK Ltd, Edinburgh
Printed and bound in Great Britain by CPI Group (UK) Ltd, Croydon CR0 4YY

Papers used by Constable are from well-managed
forests and other responsible sources.

Constable
An imprint of
Little, Brown Book Group
Carmelite House
50 Victoria Embankment
London EC4Y 0DZ

An Hachette UK Company
www.hachette.co.uk

www.littlebrown.co.uk

My daughter, Sally Mellor, kindly agreed to illustrate this book and I would like to thank her warmly for all her work.

Some of the people and places described in **The Sheep Stell** *have been given fictitious names to safeguard their privacy.*

A stell is a round, walled pen, usually built on high ground in the Scottish Border hills, for sheep to shelter from rain, wind and snow.

INTRODUCTION

A little before the current surge in nature writing, Janet White wrote a book that owes nothing to literary fashion. *The Sheep Stell*, instead, springs from a desire to record a life of passion for the natural world, a story not of travelling through a landscape but of inhabiting and working it.

In gentle, unaffected prose she describes with meticulous knowledge the qualities of a terrain, the flora and wildlife it holds, the flocks and cattle it might sustain. Her voice is the sensitive but down-to-earth vehicle for profound attachments, and sounds with clarity, precision and a faint, underlying melancholy.

Very early she decided on a future in farming: a determination held fast against the expectations of a middle-class academic family in the late 1940s. 'No one', she writes, 'could have been more thrilled to become a dairymaid.' Her companions, ideally, would be not humans but animals, and the devotion was lifelong. As a young shepherd in Scotland she tries to domesticate fox cubs; some fifty years later she is misdirecting the hunters who pursue a stag on to her property. Even the birth of her four children become integrated in the telling with those of sheep, cattle and dogs.

A craving for solitude and natural beauty enticed her successively into the life of a lone shepherd in the Cheviot Hills of Scotland, then to a remote island of New Zealand (tending her own flock of over 200 sheep), to a run-down farmhouse in the Sussex Weald and at last to a secluded hill farm in the Quantocks. So intense was her love for her New Zealand island that years later, long after she was happily married, the discovery that it was up for sale tempted her to abandon Britain altogether and return to her idyll along with an indulgent husband.

It is not hard to trace this solitary yearning back to her childhood, in which the Second World War features as an unlikely benefactor, her family evacuating to a rural haven in which the small girl grew up. She describes the bewitchment of a Cotswold valley that drew her into its private embrace: a dream of elysium never quite relinquished in adulthood.

If anything interrupts her pastoral world, it is erotic desire. An unconsummated teenage passion endures for forty years, to the end of the man's life: a passion that she rages against as a threat to her freedom. It is partly to escape this man's shadow that she departs for New Zealand, but here sexual obsession intrudes again, yet this time in the form of a rejected lover whose brutal assault aborts her island peace.

But her abiding devotion remains to the splendour and challenge of land: to its cycle of reproduction, recurring beauty and lamented deaths. *The Sheep Stell* is its testament.

Colin Thubron, 2017

PART I

SHEPHERD

CHAPTER ONE

Every animal needs its own territory and humans are no exception. My plans were clear at an early age. By the time I was fourteen I knew exactly what I wanted to do with my life. I intended to live somewhere wild and supremely beautiful. The right environment was all-important to me and I imagined searching the whole world for a place, high and remote as a sheep stell, quiet as a monastery, challenging and virginal, untouched and unknown. Perhaps I had read too many Brontë and Hardy novels for, behind a façade of scorn for everything sloppy, I was intensely romantic.

The calm domesticity of our family life had provided a surfeit of security, which developed in me a hunger for adventure and a habit of daydreaming. As I grew up, the daily routine of school became more and more irksome. Skulking in the back benches, I read under my desk cover, drew pictures of galloping ponies with streaming tails, and solitary wind-bent trees, wrote fragments of stories and poems, and drifted into idle fantasies of future freedom. If I could capture a seat near a window, I watched with envy the far-away activity in distant fields: men ploughing, trailed by gulls, mowing hay round and round diminishing islands, or stooking wigwams of golden corn. The swallows sweeping, skimming and rising above the playing field seemed favoured compared to me. It was only out of school hours that I really began to live.

Our home at that time was an old Cotswold farmhouse which we were renting until the end of the war. It was built of stone with mullioned windows, flagged floors, long passages and twisting stairs. Outside there was a sunny courtyard, an orchard of apple and perry pear trees, a barn containing a cider press, and a small pond where I kept ducks and geese.

The house was set on a natural terrace halfway up a steep spur of hill, which rose in a dome and looked out over a small, pastoral valley. It was a hard push up the lane for my sister and me with our bicycles, gas masks and homework, but if we paused to catch our breath, the unfolding view rewarded us. A bright stream winding along the valley floor, an old mill, scattered farms and cottages, a patchwork of fields divided by dry-stone walls, and all along the opposite ridge a magnificent crown of beechwoods. We loved our home and half dreaded the end of the war which would bring its loss and the ordeal of change.

On weekday mornings I got up reluctantly at the last possible moment, but during weekends and holidays all my apathy vanished. If it was fine, I put a few biscuits and an apple in my pocket and, while everyone else slept, I left the silent house. Purposefully, I walked uphill away from the village and the scattered cottages in the valley.

I had a spaniel and, glad to be out of doors, we climbed the golden hill behind our house. Golden, because nothing grazed the hill except rabbits and hares, so the grass grew long and ran to seed-heads like corn. I could lie buried and lost there on warm summer evenings, or brush a silvery slug-trail with my bare feet when the grass was softened by rain or dew. Here there were yellowhammers, pipits, and always larks, quivering and singing high overhead, then falling from the sky. Small, delicate flowers bloomed on the hill throughout the summer: yellow cinquefoil and lady's slipper, white bedstraw, blue harebells and several kinds of orchids: the early purple, the pyramid and the bee. Here, too, grew sweet-scented herbs like thyme and marjoram. It was a place to loiter in, and when I reached the summit, I always stopped and turned, for there was no better look-out point in the whole valley. The village, a cluster of cottages round the church, pub and school, slept a mile away on the opposite side of the stream. In the early morning there was little sign of life anywhere, just a curl of smoke from one or two chimneys, a herd of cows being driven in for milking, and a tractor crawling like a red ant across a green field.

Beyond the hill, an overgrown cart-track led through quiet beechwoods into a side valley. We scuffed through drifts of leaves, the dog chasing squirrels up trees. Across the path hung beaded spiders' threads, which I broke with outstretched hands and then peeled from my fingers. These cobwebs proved that no one else had walked this way since night, and I felt exhilarated by a sense of adventure. Everywhere was fresh and unsullied and I moved slowly, reverently, among the grey beech pillars.

Towards the end of the wood, I hurried forward eagerly. Very soon now I should reach the object of my journey, and taking the path which ran along the edge of the wood, with a hedge on the lower side obscuring the view, I came to a stile. Instead of climbing over the stile immediately, I would sit on the top rung and stare at the valley below. My secret valley, which I knew and loved so well, where there were no roads, no traffic, no pylons, no people, no ugliness. Just a

5

brook edged with kingcups running down from its wooded source through green and yellow cowslipped meadows, and the footpath which descended from the stile to an old, humpbacked bridge, and then zigzagged up the opposite hillside to a solitary farm.

This farm was my ideal. The grey stone house, under red-berried creeper, was small and square, approached by a flagged path between thick lavender hedges. The farm buildings were grouped around a cobbled yard, bounded by an orchard of ancient trees where a flock of white geese grazed. Above the farm rose sweeping downland pasture stippled with sheep and freedom-loving birds like swifts, lapwings, and seagulls from the Severn estuary. The contrast of lush water-meadow, breezy upland and deep dark woods gave a wonderful variety to the valley, but it was its lonely tranquillity which I loved most.

Even when I took the footpath down over the humpy bridge to cross the watercressed brook and stole quietly through the orchard and yard, the spell was not broken. Mysteriously, there was never anyone about. Sometimes an old black-and-white collie barked from a cart shed, or the geese squawked at me with outstretched necks, or I heard a rattle of buckets and cow chains from a dusky byre. It was uncanny. I felt as if I was being watched, and shut the yard gate behind me with a sense of relief that nobody had called out at my trespassing. It was not until years later that I learned the farm was run by a pair of old maids, who lived there almost like hermits and no doubt avoided me deliberately.

A grassy cart-track, steeply banked with yarrow, ox-eye daisies, and hedge parsley, took me out of sight of the farm and, dipping downhill, eventually led to a pair of old cottages. There was a well on the verge outside and in early summer usually a hen in a coop with new, yellow chicks running free. A shaggy, Old English sheepdog chained to a kennel watched me suspiciously through his doorway, nose on paws, too tired or lazy to raise his head.

A shepherd lived in one of the cottages. To my young eyes he looked about ninety with his long white beard and his iron-headed crook. We always chatted about the weather. I would have talked to him more but his dialect was so thick I could barely understand him. Often I met him driving sheep or carrying lambs in his arms and then I was envious. There was something so calm and splendid about him. He was a tall, upright man with a fine face, in spite of its many wrinkles, and he matched exactly my child's-eye picture of Jesus. Like the shepherds I saw on our family holidays in Wales, riding their ponies over the mountains, he was a figure to inspire my romantic imagination. Both he and the farm influenced my future.

Beyond the cottages the track led down to a mill-pond where moorhens called among the reeds and bulrushes. Then a path brought me back in a circle to the beechwoods and home. I seldom returned without something gleaned on the way; white violets and primroses in spring, wild strawberries in summer, hazelnuts, blackberries and mushrooms in autumn, and holly, pussy willow and catkins in winter. I loved walking and there wasn't a footpath for miles which I hadn't explored. If it was warm, I armed myself with a stick in case I should come across an adder basking in the sun. Adders and thunderstorms were my only enemies.

The war was too remote to frighten me. Nevertheless, there was no escaping its influence, with the nightly blackout, the siren, the menacing murmur of aeroplanes and sometimes the boom of a far-away bomb dropped over Gloucester. But mostly the war meant trivial hardship: a shortage of new clothes, which didn't bother me since I preferred old ones; eking out my sweet ration; paring away the weekly cube of butter; gathering rose-hips and tufts of wool from the hedges and remembering to take my gas mask to school. The horror of war was only brought home by the nine o'clock news. This news seemed to come from another planet. It was so difficult to believe that such things were actually happening: ships going down, planes never coming back, houses flattened, people gone.

As a child my faith in God was strong. But war and God did not mix. It was too cruel. Better to shut it out of my mind and concentrate on valuing the peaceful countryside. I wished that everything in the world could be fair, but if happiness had to be dealt out like a pack of badly shuffled cards, I recognised and appreciated the aces in my hand.

I often helped out on a local farm. I learnt to milk a cow, feed the new calves, groom the big cart-horses, stook sheaves of corn. I loved even the most menial jobs, like washing eggs or hosing down the cowshed. I received no money for my efforts, but if I was lucky, I was sometimes allowed to exercise the farmer's lively 16-hand mare. She was a challenge I could not resist, and much more exciting than the lethargic riding-school ponies which could be hired for two-and-sixpence an hour.

I couldn't pass an empty house without trying the doors and windows to explore inside. I enjoyed looking in country churches and if no one was about, I strummed on the organ or tolled the bell to find out how it sounded. High walls invited curiosity, and summer-houses and lonely barns were the best places to shelter from rain. As for the lure of water, I knew every pond and lake for miles and where to go to skate in winter, swim in summer, or enjoy a little furtive punting or rowing in some ancient skiff under a welcome canopy of overhanging trees. Sometimes I was caught red-handed, sometimes I took to my heels at the sight of a distant arm-waving figure, but nothing deterred me from my deliberate search for fresh places. And when I found somewhere like my secret valley, I returned again and again as if my senses could never be satisfied. I wished more than anything that I could paint, or write poetry, for there seemed no other means of celebrating such loveliness. It was no good telling people and taking them there, because to go accompanied meant talking, and talking meant that the birds flew away, the flowers were trodden, the view unseen and the essential rapture was lost. So more and more I explored alone.

The war ended when I was fifteen. I remember Victory Night because I climbed the little mountain behind our house to watch triumphant beacons blazing on the distant hilltops. Everyone was celebrating. It was a beautiful evening, with a velvet black sky, bright stars winking and the moon glowing yellow over the wakeful villages and towns. No more shuttering of lighted windows and blackout precautions. But the end of the war meant the end of our rural life.

My father would soon be working in London and we would have to leave our Cotswold home. The ducks and geese would have to be sold. The dog would have to learn to walk on a lead. And all those familiar, lovely places would be left behind, perhaps for ever. My sister would be going to Cambridge in the autumn, so I should have to start at a new school alone. Some primitive animal instinct attached me to my own particular territory, and I was panic-stricken at the thought of being torn away.

Back in my own room that night I could not sleep for wondering what the future held. That room with its latch door, wide stone window-sill and cupboard crammed with a squirrel hoard of notebooks, diaries and drawings, had been my bolt-hole for the growing-up phase of my life. I had been happy here, knowing no suffering beyond the death of a pet, dull days at school and the mutterings of a war so absent from our valley as to seem unreal. Now I was afraid my luck could not hold.

CHAPTER TWO

My parents compromised over where to live and found a house in a large Chiltern village within daily reach of London. It was a pleasant enough modern house, but nobody's ideal. I tried to ignore the distant factory sirens, the rumbling trains and the click of suburban gates along our laurel- and privet-hedged road.

There were fields at the end of the garden and farms and footpaths for the finding. Sometimes I walked along a towpath beside a canal where there were swans and stubby willows but the water smelt of factory effluent. The paths were worn and dirty and no matter where I wandered, I could find no wilderness anywhere. I was miserably homesick and hated my new school.

Gradually my restlessness turned into downright rebellion. I wanted to leave school and get a job on a farm. I had already gained the necessary examinations to give me a place at a university if I wished. Since I did not wish, I argued that there was no point in staying on for further study. Why shouldn't I educate myself by reading and working at the subjects which interested me?

At home I threatened to run away. Realising how I fretted, my parents eventually agreed to let me apply for a job on a local farm. No doubt they hoped that a few weeks of rising at six o'clock in the morning would cure me, especially as it was January and I would have to cycle to work in darkness and all weathers. At least they couldn't accuse me of financial ambition for I should earn less than two pounds a week for my efforts. However, these drawbacks were just the right kindling for defiance. No one could have been more thrilled to become a dairymaid.

I loved the smell of cows and hay and malty brewer's grains; the clang of buckets and drumming milk and the whistling of the men. Each cow was named and the head cowman would call, 'Come along, little maid' to the nervous heifers and the newly calved. My special Sunday perk was a carton of thick yellow cream. Even the return journey was a pleasure because it was nearly all downhill and just when I was tired, I could freewheel home effortlessly.

Though much of the work on the farm was routine and repetitive, for the moment I was content. On arrival in the morning, my first job was to set out the sterilised buckets and stools for the milkers and prepare the cooler, churns and separator for the incoming milk. Everything in the dairy was kept spotlessly clean.

In winter the cows remained tied and bedded in the adjoining shed and when their udders had been washed and their food bowls filled with measured concentrates, the men put on their white caps and coats and came to the dairy door to collect their buckets. The whole herd of about seventy cows was milked by hand.

The milking team consisted of the head cowman, the second and third cowmen, two boys, two other girls and myself. I could only milk a couple of cows before the rush of incoming milk meant that I had to help in the dairy. Milking time was the best part of the day for me. I loved to hear the rhythmical drumming of milk into buckets and to see my cow munching her cake with relish. My wrists and arm muscles ached but I knew that only practice would enable me to achieve that glorious bubbly froth in my pail like the cowmen.

At the dairy door the yield of each cow was weighed on a spring balance and recorded. Then I poured the milk into the cooler where it rippled down over a water-chilled metal panel and through a strainer into a churn. A few buckets of milk went into the separator which whirled the lighter cream to the top so that it could be tapped into cartons while the thin blue skim milk was run off into pails for the pigs.

After milking, all the dairy utensils had to be washed and stacked in the sterilising cabinet in which hot steam destroyed any harmful bacteria. Finally, the walls and floor of the dairy itself were swilled down and woe betide any of the men who came in with muck or hayseeds on their boots! The rest of the morning was occupied with feeding and bedding calves, slicing mangolds, grinding oats or bagging corn in the dusty granary. In the afternoon the cows were fed and milked again, then strawed down for the night.

When the days lengthened and the grass started to grow, the cattle were turned out into the fields and only driven in twice a day for milking. I liked being sent to fetch the herd. The old knowing cows queued at the gate and eventually even the most reluctant heifers followed, but there was never any hurrying. We took our time and I was glad to be out of doors. Any field work appealed to me. I even enjoyed the more punishing jobs like hoeing and singling roots, cutting out the weeds and surplus plants with the blade of the hoe to leave a neat evenly spaced line of strong healthy seedlings. But the height of bliss was to ride in one of the carts or

13

haywains behind the big Percheron horses which were kept on the farm. I envied the carter. I would much rather have been out with the horses all day than working in the dairy.

I stayed at the dairy farm for a year. My parents gave up all hope that early hours and manual labour would cure me of the desire for a farming life. They reminded me that it was not too late to change plans and go to a university, but my mind was obstinately made up. I wanted to widen my experience so I decided to leave home and take a job which would enable me to live on a farm. Like a fledgling testing its wings, I ventured further and further afield.

My first long flight took me to Scotland. I journeyed into a vacuum, knowing nothing of the place or people beyond what I had gleaned from a brief exchange of letters. The farm was situated near the shore of a small loch and surrounded by high, open hills with grassy flanks and heathery crowns. Clear burns tumbled down the gullies in a series of waterfalls and limpid rock-lined pools. Curlews, blackcock, ravens and grouse were all familiar here, and everywhere sheep grazed: Blackfaces with curving horns and long coarse wool, nibbling the close turf beside the narrow empty roads and browsing among the heather shoots on the high tops. Their half-grown lambs were noisy, losing their mothers and bleating for them in the fern and outcrops of rock. A plantation of fir trees sheltered the steading and a few small fields, enclosed by stone dykes, lay round the square, grey farmhouse. A house which reminded me of my Cotswold ideal, though this Scottish home was much more austere and windswept.

The property had been tenanted by the same family for five generations and the Master and Mistress, as they were called, were liked and respected by everyone in the district. They had one son aged twenty. He was tall, lithe, black-haired, with large dark eyes and a wide ready-to-smile

14

mouth. I admired him immediately, especially the effortless way he could vault a five-barred gate and the energy with which he worked and sang at the top of his voice. He always spoke his mind, and although I shrank from his searing criticism, I valued it for being so truthful.

When he said, 'That's a terrible-looking dress you're wearing!' I felt like ripping it to shreds and throwing it at once into the dustbin. When he told me I was ignorant because I couldn't help him finish his crossword, I felt ashamed and sure that he was right. And when he mocked me for my English accent, I hardly dared to open my mouth.

A gigantic sense of inferiority plagued me and was not helped by the efficiency and popularity of the resident land-girl. She was pleasant and intelligent and, being six years older than me, had gained considerably more social poise and experience of farm work. Beside her, I felt naïve and useless, and green with envy when she and the farmer's son went off together on some tractor job requiring two people. Too often I seemed to be the odd one out, left behind to peel the tatties or churn the butter.

However, there were so many new activities to absorb my attention that I didn't waste too much time brooding on my shortcomings. The summer was hot and dry and across a couple of fields from the farmhouse a river ran out to the loch, providing a perfect dooking pool for anyone hardy enough to keep the salmon and trout company. I loved to swim and there was no better place than this cold invigorating water to cool off after a day rolling fleeces or forking hay. To the lean, wiry hill-shepherds working on the farm, I was a soft, lazy southerner. They teased me, especially when I made foolish mistakes, like driving the tractor round a field while the roller I thought I was towing had come adrift and slid to a halt one hundred yards behind me. Or when, paring sheeps' hooves with a penknife, I sliced my own hand. Or when I waited in vain at four o'clock in the morning to go with the shepherds to gather sheep for the clipping, unaware that the thick blanket of mist on the hills meant a reprieve in bed.

Everything was new to me. The sheep were washed in a river pool to remove grit, dirt and grease from the fleeces before shearing. The method of shearing was different too. We would sit on low benches in front of a foot-bound ewe, snipping away with pointed handblades which had been rubbed razor sharp on a whetstone. The fleece was then trimmed of any soiled dags, rolled into a bundle from the tail end to the neck and packed into a big hessian sack slung like a hammock between two poles. Wool is springy and takes up too much space so, as the bag got full, it was necessary for someone, usually me, to climb inside and tread down the fleeces, floundering and hanging on to a rope to keep my balance. When the bag bulged to the brim with wool it was sewn up with a gigantic curved needle and string.

Haymaking was different too in these northern hills. A team of us went out with scythes to mow round the perimeter of the hay meadow so that not a blade of precious grass should be wasted. The Master showed me how to hold my scythe and swing it with a wide, level, regular sweep which heaped the cut grass and clover to one side. As soon as there was space for a tractor, the rest of the crop was cut by a finger-bar mower. When the grass had wilted for a day or two we turned the edges of the field with hay-rakes while a horse-drawn implement flipped over the rest. Sometimes the hay had to be turned again and again, tossed and fluffed up with pitchforks to let the air into it. Then we built pikes or little conical stacks which had to be netted down to withstand the high winds. When completely dried out, they were transported to the barn by winching them on to a bogie.

Each day there were also routine jobs like milking Juliana the house cow, feeding the sheepdogs, moving the coops of cluckers with their newly hatched chicks on to fresh grass and chasing Macgregor, the pet lamb, out of the garden.

Though sceptical of my skill as a worker, the shepherds were always willing to answer my questions. The Master, as he explained himself, was 'a wee bit dull in the lugs', which impeded conversation, but he was aware of my interest in sheep-farming and sometimes took me to shows and sales. His son and the land-girl belonged to the local Young Farmers' Club, so the three of us went to meetings together and to the shearing and stock-judging competitions which took place that summer. I was even persuaded to go to my first dance, held in a small wooden hall on the banks of the loch. An accordian played Scottish reels and polkas and as the dancers grew warm they threw aside their coats, collars and ties. The floor shook beneath their pounding feet and I had to watch my toes, for some of the shepherds still wore their heavy tackety hill-climbing boots. I didn't know the steps, but there was no shortage of partners willing to teach me and my evening was made when the farmer's son said as he danced with me, 'That's the first kind o' decent dress I've seen you in!' Such approval was sweet praise indeed.

There was a gamekeeper on the estate, an eccentric man who had been everywhere and done everything, or so he said. One day he took me panning for gold. We spent hours wading up and down an icy burn with frying pans and sieves, sifting handfuls of sand and gravel collected from the pools. Finally, he pointed out a speck of yellow in his pan. That, he assured me, was gold! I still don't know if he was right, or simply determined not to disappoint me.

A small fat pony was kept on the farm, and since nobody else seemed to want to ride her, I exercised exclusive rights. There was nothing I liked better than exploring the surrounding country on her. Range upon range of hills beckoned in every direction with the sound of fast-flowing water, the glorious piping of the curlew and a special peaty-heathery tang to the air. Sometimes on a Sunday when the others went to the kirk, I took a picnic and a map and spent all day alone on the hills.

The most beautiful place of all was a remote tarn, two thousand feet up on the tops, away from all roads and habitation. I remember swimming there and seeing the shale, water-weed and small, freckled trout as clearly as through glass, while overhead ravens croaked and from the shore astonished sheep stood and stared. If paradise existed, it was surely here, where sky, water, grass and stone were all flung together in perfect colours and pure lines. The sheep-tracks lured me, twisting among dark peat hags, over grey screes, skirting treacherous green moss and white-tufted cotton-grass, then plunging into purple seas of heather. Here, I could walk or ride all day and never meet a soul.

That summer was so much to my taste that I dreaded the day when I should have to leave. But in the autumn I was due to enrol as a student at a county farm college. Months before, I had applied to take a year's theoretical course in agriculture and so I knew that my stay in Scotland was limited.

The morning the Master drove me to the station, I sat very quietly, struggling with a choky feeling in my throat. But one thing was clear to me. There would come a time when I would be a shepherd and return to these hills.

CHAPTER THREE

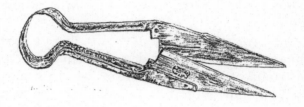

I had not been at college a fortnight when something natural, but completely unexpected happened. I fell passionately in love. Thrust into a lecture room full of young men with only one other girl to keep me company, I suppose it was not surprising. Once I had seen him, I could not turn away. I was hypnotised, addicted and furious with myself. Such infatuation was a threat to my freedom, self-sufficiency and ambition. The last thing I wanted was to become dependent upon anyone else and veer off the course I had chosen. I raged at my weakness, but no matter what fine resolutions I might make when safely alone, as soon as I was confronted by him, I was lost.

The lectures were thrown away on me. Past and future were forgotten. My life was monopolised by his movements. I could recognise the particular roar of his motorbike, his heavy step and the scrape of his chair. Perhaps this obsession would have wilted if my awareness of him had been entirely one-sided, but it was not. On the very first morning, his head turned to register my arrival, and when in lectures he chose to sit several rows in front of me, he made a point of lounging sideways so that it was possible to glance around. Over the months, with gathering momentum and excitement, our eyes converged, mine unwilling and hostile, his winking and smiling.

I watched him like a hawk, ready to fall upon his faults and crush out of existence all delusions, but it was no use. If imperfections existed, I could always find excuses or reasons for them. Meanwhile I earned a reputation for being aloof. I would go out with no one. I was indifferent to all other men.

Not even my dreams could have supplied the ending to that year at college. On the last day of the final term he came with an armful of roses for me. We swam in a nearby river and lazed and talked on the bank all afternoon in the hot July sun. At a ball that evening we danced together until the early hours and returned to the cool, dark river to swim again. A crescent moon shone as we walked back across damp meadows scented with the aftermath of hay, while our locked fingers proved that there could be no greater rapture in life than to touch someone for the first time with innocent unspoken affection.

On leaving college, I never expected to see him again. We had not mentioned the future and because I was young and ambitious, I could accept this fairy-tale ending as the right and perfect one. I set off for Ireland to live a hippy type of existence, hitch-hiking and sleeping rough in hay barns and derelict cottages. A month later, back in England, I found his letter waiting for me, the handwriting unfamiliar then, but to become printed indelibly in my mind. I replied at once. A fertile correspondence and devoted friendship began, which survived both his marriage the following year, and my own eight years later. I came to

20

know and love not just the outer husk of him, but the inner core. Our lives ran parallel for many years with our respective farms, sheep, four children apiece and shared passion for the country. But I loved him too much and ended up exiled from his company.

When I returned from Ireland, I began to look for a job. I was now eighteen, feeling restless and unsettled and dithering about where to go next. My advertisement for work on a sheep-farm brought in a shoal of replies. One of them intrigued me. It was a semi-literate letter written in a round, childish hand with fantastic flourishes added to every looped and capital letter, while under the proud signature was drawn a tremendous double zigzag. 'The farm lies high,' said the letter. 'We keep a thousand sheep. If you come, you would be one of we.' The farmer suggested I went for an interview and so that was what I did.

A train heading west along a single-line track took me from the Midlands into Wales. Big shallow-roofed Georgian houses and black-and-white Tudor cottages were replaced by plainer buildings, grey or whitewashed and blue-slated. Meadows, orchards and trim thorn hedges gave way to steeper pastures, dry-stone walls, forests of uncurling bracken and fast-flowing brooks. Each station seemed smaller and emptier than the last, and when the train wheels slackened for my destination, there was no sign of any town or village, just a patchwork of rush- and boulder-strewn fields and small spinneys climbing up the valley sides. I tugged at the leather strap which let down the carriage window and waited apprehensively, ready to open the door. The station was no more than a deserted, wooden platform and a tin hut. No other passengers got off. I looked round anxiously, expecting to see a waiting stranger, but the only man in sight was the blue-uniformed official, an ancient sharp-nosed mole of a man, who emerged from his dim

21

shack to collect my ticket and retrieve the mail-bag thrown out by the guard. I was eyed suspiciously and branded as a foreigner.

Passing into the station yard, I looked hopefully for a car, but the square of asphalt lay empty. The train had arrived a little late, which was all the more reason for expecting the farmer to be there first. It was an anticlimax to find no one to greet me and when at last I heard a vehicle, my disappointment was all the greater when it turned out to be only a red postal van come to collect the mail-bag. Soon after this, the little bent mole began to bolt and bar the entrance to the station.

'What time is the next down train, please?' I asked desperately. He looked up in surprise. 'There's not another today,' he said as though I were being quite unreasonable. 'Ten-thirty tomorrow morning she comes.'

Dismayed, I asked how far it was to the farm.

'It's a long six miles,' he assured me. 'Maybe seven and that's taking the shortcut up by the bog.' There seemed to be no alternative but to walk, so I set off in the direction he pointed out.

After the stale-smoke smell of the railway carriage, it was pleasant to be out of doors walking. The road climbed steadily towards a high, bare skyline and with the gradual rising of the land the trees became stunted and moulded by the prevailing wind. Harebells, scabious, furze and broom predominated along the wayside and the breeze, which had been imperceptible in the valley, now made the tall summer grass wave and bow. Eventually I branched off the road along a cart-track, deeply rutted by tractor wheels and imprinted by pony hooves along the central spine of turf. This led to a farm where I asked the way again and was directed through a wicket gate and told to cross over the mountain, pass the bog and head for the only visible trees.

22

Misgivings nibbled at me. Perhaps it had been foolish to come. Maybe the farmer had already found someone for the job and the interview was no longer necessary. Besides, there would be no pleasure in working for a man who was too slack and unreliable to remember to meet me. Nevertheless, to turn tail now would be to let him off completely and I was resolved that he should pay for his neglect. At least he could provide me with a square meal, a bed and a lift back to the station in the morning.

Continuing along the path with new determination, I came to the bog. Though boulders had been placed strategically in the wettest hollows, my shoes became muddy and my legs splashed with peat-brown water. A scattering of mountain sheep grazed among pools and sedges and, coming over a knoll to drier ground, a group of rough-coated ponies and foals wheeled away, long manes and tails flying. Then I saw the dark blue-green tips of firs peeping from a fold in the mountainside and a thread of smoke drifting from a hidden chimney. The farm was like an outpost in the hills, remote from roads and houses, isolated in a setting of open, windswept beauty.

A file of newly shorn sheep, white, naked and goatlike, emerged along the green brake which led through the fir plantation. They stared with brazen yellow-brown eyes then, seized with modesty, retreated to the trees and bracken, bleating to their lambs. The farmhouse was hidden from sight until I came to the end of the wood. Then I stood rooted to the spot in horror.

The place was ugly. A slum. The rambling, stone house had been painted a lurid red, an obscene red, a red no one in their right mind would paint any building in the hills. Every single window was shut, but the back door stood open and a scruffy moulting hen skulked on the doorstep. A pig wallowed in the mud outside the garden gate, which tippled from a broken hinge. A line of murky washing flapped

23

languidly between two ancient apple trees, while a horse rubbed himself against the forked prop. There was no garden, but a patch of bare earth, soiled with chicken droppings and pressed into a desert by flat webbed feet. Near the unkempt buildings, a binder, a hay-rake and a set of harrows lay rusting and almost camouflaged by nettles.

I was on the point of retreating when I was seen by a gander and a wall-eyed collie. The sudden screeching of geese and the barking and growling of the dog brought a girl to the porch, leaving me with no alternative but to go forward.

The girl seemed surprised to see me. 'We didn't think you'd get here,' she said without explanation. 'My brother's shearing. Come and meet him.' She was dressed in a shrunken faded dress and grimy tennis shoes and splashed heedlessly through the puddles as she led the way across the muddy yard. We came to the open doors of a large barn. 'Ivor! Look who's here,' she called.

A gigantic figure loomed out of the dim interior and a hand wrung mine in a finger-crushing grip. We stared at each other, astonished and unprepared. He was a massive, handsome man with curly red hair and a bushy red moustache. Eyes, too blue and twinkly to look honest, were half-shut in the glare of the sun, and when he began to smile, his cheek muscles tightened. His clothes were old and shabby, boots laced with binder twine, trousers shiny at the knees from contact with greasy wool, and his shirt was ripped. A ginger stubble, several days old, grew on his jaw, his knuckles were scarred and a thumbnail blackened.

I felt shy of him. His eyes were too shrewd, too impudent.

He asked how I had reached the farm and I told him that I had walked.

'All that way? By yourself! We never thought you'd come,' he said.

'But why not? Didn't you get my letter?'

'Yes. I was coming to meet you,' he assured me. 'I should have done if it hadn't rained yesterday. We'd have finished shearing, but the rain held us up. A couple of neighbours are helping, but they can't come over tomorrow, so we're punishing ourselves to get done tonight.' Then he turned to his sister. 'Make a cup of tea, Sue. I'll tell the boys. They could do with a break.'

He beckoned me to a gateway a few yards from the barn. 'There!' he said proudly. 'You can see most of the farm from here.'

A green meadow sloped down to a small reed-bound mere, beyond which the land climbed again to a high tree-less crest. The valley twisted away to the left where distant hump-backed mountains lay like dark storm-clouds on the horizon. To the right a trout stream and its quicksilver tributaries rushed down from the open hill, a stunted thorn and rowan or two clinging to the bank.

Ivor began to point out the boundaries, the two enclosed fields shut up for hay, a pocket handkerchief of land on the valley floor planted with oats and potatoes, the sheep-dip and folds, an old stone building where outlying cattle could shelter and the mountainside, acre upon acre of it. 'There's space here,' he said. 'I love space.' He kept looking round at me while he talked and suddenly asked my age.

'I imagined you'd be much older, I don't know why,' he confessed. Then, 'Can you milk a cow?'

'Yes.'

'Drive a tractor?'

'Yes.'

'Ride a horse?'

'Yes.'

He bent down, picked a grass and held it out to me. 'What's that?'

'Meadow fescue.'

'Can you rise early?'

'If I have to.'

He began to laugh. 'I know one thing you can't do,' he gloated. 'Shear a sheep!'

When I said I could, he didn't believe me. 'Let's see you,' he demanded and swung round towards the barn.

The men broke off their conversation as we entered the barn and I was uncomfortably aware of them pausing to stare when Ivor handed me a pair of shears and went to catch a ewe from the pen. He hesitated before handing the sheep over. 'You'll make your dress dirty. You must be tired, walking all that way.' But I wanted to show him what I could do.

I didn't make a very good job of the sheep with him standing watching. Twice I cut the ewe, not badly, but nicks in the skin which bled sufficiently for noticeable stains to spread like red ink on blotting paper over the sheep's white body. I swore and heard the men laugh. When the fleece finally slipped to the floor, Ivor leant forward and branded his initials on the animal's flank, I let her go and we stood in the doorway watching the ewe retreat. A lock of wool still fluttered from her tail and I felt cross not to have shorn her better, but he said, 'You were quick. Harry hasn't finished yet and he began before you.'

As we walked towards the house, he told me the job was mine if I wanted it, but I said I needed to think it over.

We came to a large, dim kitchen with low, beamed ceiling and stone-flagged floor. The big central table was laid for six, with a red hunk of corned beef, an enormous, yellow triangle of cheese and a loaf of bread. Through the window, I could see the men washing in a horse trough across the yard, but I was offered a bowl of water in the scullery. A cat was cuffed off the table and when the men came in, we all sat down to eat. Sue got up from time to time to refill our cups from the big, brown, enamel teapot keeping warm on the range.

A constraint fell over the company and I suspected that I, as a foreigner, was the cause. Once or twice I glanced round and caught a furtive eye which was rapidly averted. Only Ivor seemed unperturbed, talking casually, watching my plate and offering me food. I had lost all sense of time and stared in disbelief at a grandfather clock in the corner of the room, which said twenty minutes past eight. As the men thawed out they began to discuss the shearing, their individual tallies, the sand and grit in the wool and the weight of the fleeces. But there was one man who kept silent. I noticed his ginger military-looking moustache and later I learnt that he was Ivor's brother, but apparently they were scarcely on speaking terms. That night Sue told me that her brothers had divided the farm, the stock and the machinery in two halves and could hardly bear to lend one another an implement. However, for the time being I was unaware of this feud and watched as Ivor filled his pipe with tea-leaves from the caddy for an after-supper smoke.

The hours of daylight shrank and there were still sheep to be shorn, so I offered to help. At first he refused but eventually Ivor said that if I didn't feel too tired I could help him mark the hoggets.

While the men returned to the barn to finish shearing, we crossed a cobbled yard to a range of old stables where a mob of shorn sheep were tightly penned. Ivor decided that I had better do the marking and open the door, while he man-handled the animals. A long iron, with his initials I.W. on the end, had to be dipped into a can of red marking fluid and pressed on to each sheep's flank. My first attempt was upside down, the next few looked blurred and indistinct, but eventually I got the knack of pushing the iron down quickly but firmly. There was an overpowering sweaty

27

smell of sheep in the building and a continual jostling and scrabbling of hooves on the uneven floor. Already the light was dwindling and through the open top half of the stable door, I could see bats zigzagging above the yard and the first stars winking in a clear sky. Behind a tall pear tree, a huge butter-yellow moon was rising to act as our lantern.

Once a ewe struggled so much while being branded that the paint on its flank smeared my skirt. Ivor was very concerned. 'Your best dress spoilt! I should never have let you work.' He would not believe me when I said it was only an old cotton dress and unimportant.

When there was more space, the sheep became increasingly difficult to catch, bolting to left and right and dodging under the mangers. Ivor paused for breath more often and rubbed his back. Sometimes in the semi-darkness our hands clashed on the latch of the door, or our shoulders bumped as we bent over an animal.

'Do you think you'd be happy living here?' he asked suddenly.

'It's the right kind of work and place,' I began.

'But you're not so sure about the people?'

'I never said that.'

He laughed. 'I could tease you,' he said. 'I love teasing. You're not a bit like I expected. You're just a child, really.' And then, 'I wouldn't blame you if you turned the job down. But we don't work like this every night. The moon's only bright enough occasionally.'

Eventually the last sheep was released and we leaned in the doorway. 'God I'm tired!' he said, but his tone was filled with the satisfaction of seeing a good job done. 'I'd not have managed without you. I'm very grateful, indeed I am.'

'I've enjoyed it,' I responded with honesty.

'The evening's been the best part of the day,' he reflected. 'Come now. We'll have a drink of tea and then you can sleep till dinner time if you like. I've a mind to milk the cow before I go to bed.'

Outside the sheep had vanished into the night, but there was still a persistent bleating in the distance, as ewes hunted for their lambs. The moon, now brilliant white instead of yellow, had risen high in the sky and threw mysterious shadows.

We said goodnight in the kitchen, then Sue led the way upstairs, carrying a lighted candle. I discovered we were to share not only a bedroom but the large brass bedstead. There was no bathroom, only a slop pail and a jug of cold water, full of drowned flies. The room was shabby with peeling plaster, blue patches of mould, furniture pitted by woodworm and a small low window wedged with a piece of cardboard to stop it rattling in the wind. Sue pulled her dress over her head, dropped it carelessly on the floor, flipped back the murky, flannelette top sheet and promptly got into bed. I could feel her curious eyes watching as I cleaned my teeth, washed my face and hands, unpacked and finally put on my nightdress. I was glad to snuff out the candle, but climbed reluctantly into the double bed with its sagging mattress rolling us together and the stale, tickling sheets.

Somewhere out on the hill, a lamb bleated incessantly for its mother and a dog in the yard barked in frenzied retaliation at the weird yowl of a vixen.

'Damn foxes! They killed nine of my pullets on Tuesday,' Sue said, scratching away like a hen herself, until I too began to itch and fidget. Inquisitive and relaxed, she seemed more disposed to chatter than sleep.

'I wouldn't like Ivor for a boss,' she admitted. 'He works all hours. And farming's tough for a girl. Still, I hope you'll stay.'

Any moment now she would ask outright if I was going to take the job, so I yawned noisily and suggested we should

29

get some sleep. Sue pulled a lion's share of blankets round her in a cocoon and within minutes was snoring softly. But I lay awake a long time wondering whether I could endure the sordid sheets, the mice – or were they rats – scrabbling behind the skirting board, the drowned flies and endless corned beef, all for the sake of hills and sheep and a man with flaming red hair.

In the morning a thick, grey mist obscured the landscape. After breakfasting on bread, very salty fatty rashers of home-smoked bacon and tea into which wood ash had fallen, I borrowed a pair of gumboots and an old raincoat and followed Ivor outside. He dragged open a sagging five-barred gate and stood aside for me to pass through.

'Well, you'll not fall in love with the place today,' he warned me with a wry smile. 'You'll see it at its worst.'

A lot of rain had fallen in the latter half of the night and our boots squelched in the sodden ground as we tramped across the poor pastures where rush, boulders, fern and coarse, moorland grasses predominated. The mountains were lost in cloud, the mere lay smooth, grey and vacant, and the newly shorn sheep looked cold and pale. Three collies followed behind us, routing rabbits and small birds, stonechats and linnets, out from bracken and whins. We came to a band of bullocks and halted.

'What do you think of them?'

'They look fit.'

'I shall have to sell them before winter. Nowhere to house them. Nothing to feed them.'

Gradually I was beginning to realise the poverty of the farm, and that all Ivor's fine plans for reclamation, new fences, new buildings, more stock and more land were dreams which might never materialise. These steep, hard, thin-soiled hills offered little scope for ambition. No wonder his boots were laced with binder twine and he offered me a pittance of a wage. But I could have a pony to ride round the sheep he said, and he'd get me a collie dog.

We waited under a hunchbacked thorn tree for the rain to ease. There was nothing to see. We were shut in by the mist which hung in beads on the fronds of Ivor's rakish moustache and his bushy, red eyebrows. At last he asked the question I was dreading. Did I want the job? Had I made up my mind?

'You're not sure', he said, observing my hesitation. 'There's a snag.'

I couldn't deny it. Such conflicting wishes churned in my mind. But how could I tell him his house was too dirty, the food unpalatable, the whole atmosphere of the place overwhelming and claustrophobic.

'I don't know how to tell you,' I confessed.

'Shall I guess? We're too rough for you. Isn't that it?' And when I was silent, he added, 'Indeed, I don't think you ought to come here.'

He wanted me to take the job, he said, but he was concerned for my happiness. He had a bugbear too, he admitted, but he couldn't tell me what it was either. Not yet, not now, he really couldn't.

I was consumed with curiosity.

The mist showed no signs of lifting and there was no point in going on. We started to walk back slowly by a different path which crossed and recrossed a mountain stream. Ivor led the way, thrashing back the brambles and thistles with his stick and offering an outstretched hand over the slippery log bridges and stepping stones. All the way, he cross-questioned me, not only about farming, but about what I believed and wanted and thought. No one had ever put me through the hoop so thoroughly in so short a time.

In the evening, he tested my knowledge from an encyclopaedia, the only book visible in the house. Next morning,

31

on the first stage to the station, it was my nerve he tested as he drove the tractor down a precipitous slope, laughing over his shoulder while I stood on the drawbar, clinging desperately to the mudguard. Then, reaching the old shed where he kept his car, he made me reverse the car down the road. I backed too far into the hedge and scratched the paint, but he seemed amused rather than angry. Finally we arrived at the station, three-quarters of an hour too early.

While we sat in the car waiting, he smoked one cigarette after another. I urged him to tell me his bugbear, as he called it. He struggled with embarrassment. 'The truth is,' he said at last, 'you're far too attractive. There now, I've told you everything. If you were old, or coarse, or ugly, it would be different, but we'd be living in the same house, you see . . .'

In my ignorance, I did not know how hard proximity could be for a man. Besides, I was already in love and secure.

But when the train arrived at the platform and I was saying goodbye through the carriage window, I understood better. He tried to push some money into my hand. 'I don't want it,' I protested.

'Yes, you take it! For your fare and all that work.' And he flung it in at the window as the train pulled away.

Alone in the compartment, I picked the notes up off the floor and sat crushing them in my lap and battling against tears, as I felt the sudden, unexpected tug and finality of our parting.

I turned the job down, but regretted it ever afterwards. What miserable cowardice and weakness to be so concerned with material comfort when the landscape was so much to my liking. Was it really the flea-ridden bed which weighed the balance? Or was I deceiving myself and flying from a situation which would surely have become impossible? I could not be certain of anything except that I was tantalised to know what would have happened. The whole episode was like a book half-read with the crucial end chapters missing. Did the brothers continue to fight over the farm and eat at the same table in obstinate silence? Would Ivor achieve his dreams or sink into poverty? One day a letter arrived, all loops and proud flourishes. He referred to the night we marked the hoggets. 'I have tasted a piece of heaven,' he wrote. But further down the page he added. 'I think you will not come.'

CHAPTER FOUR

More than a year passed before I set off for Scotland to work as a shepherd in the Cheviot Hills. This time no gnawing doubts troubled me. As soon as I set eyes on the rolling heather-crowned ranges flecked with sheep, and the lost and forgotten cottage where I would live, I knew I had made the right decision.

It could not have been the primitive life, nor the isolation, which had deterred me in Wales, for here was a place humbler than Ivor's farmhouse, uninhabited for years and even more remote. The cottage stood near the head of a burn, two miles from any road or house, reached up a winding valley by a rough path which forded the water five times. On every side rose steep, open hills with the great, high dome of Cheviot itself, sometimes shrouded in mist, sometimes dark and near, on the skyline. This was a wild landscape, full of blending, variegated colours: green, yellow and rust-red on the lower flanks then, above the fern line, purple, grey, blue and black, the shades changing with the seasons and brightened, or softened, by the light. There were no trees except a big, protective ash standing guard over the cottage and a few rowans overhanging the small trout pools, where in summer cascades of honeysuckle grew from the damp clefts in the rocks.

The stone cottage was very simple, with four rooms, a central porch, a window to each side and two dormer windows above. Inside, it was crudely equipped for a shepherd at lambing time, with an iron bedstead, kitchen table, a hard chair or two, an old cooking range and a cold-water pump. A few steps from the door, the burn tumbled over a ladder of boulders into a thigh-deep pool, and to the right stood a small byre, stable and sheepfold, all set within a tiny dry-walled meadow. I borrowed a horse to keep in the meadow and to carry my stores up the valley once a week. And I bought a Border collie called Loch to work the sheep. At last my ambition was fulfilled. Here was the work and environment I craved, and the freedom which I had lacked but needed and valued so much.

A measure of solitude had become essential to me. I wanted time to think and appreciate fully the ideals I had formed and found. Concentration was only diluted by company, but I learned very soon that it was impossible to remain entirely self-contained and isolated. Slowly I was drawn into the small, tight community of shepherds known locally as herds. The day-to-day herding, which entailed driving the sheep down from the high tops in the early morning to graze the better valley pasture and sending them back up to their hill camps at night, I performed alone, but there were seasons of communal work when neighbouring herds helped each other to gather, clip or dip their flocks, to make hay, cut peat, burn heather or mend boundary walls.

Locally these walls are called dykes and repairing them was a task which we could tackle whenever there was time to spare. The agile Cheviot and Blackface sheep were the chief culprits for making gaps but sometimes walkers, climbing over the walls, sent loose stones tumbling. If we were mending a long length of dyke, guide strings were tied on either side to keep the building work straight. Stones were carefully selected to fit together in double thickness, nearly three feet wide at the base. The largest boulders, known as *throughs*, were set across the centre to act as ties. My job was to pack the core with smaller stones and find suitably thin cope-stones to be placed on edge along the top. Dyking was satisfying but slow work.

More exciting was the heather burning. Selected stands of old woody heather were burnt in the late winter to encourage new young growth in the spring for sheep feed. A quiet dry day was needed with just a gentle steady breeze to fan the fire in the right direction. Then two or three of the herds would go ahead to set the heather on fire with lighted brands made from oil-soaked rags bound tightly on to wire. The rest of us, carrying fire-beaters, would follow, skirting the charred area and beating out the tongues of flame and smouldering patches which threatened to spread too far. Sometimes we caught glimpses of snakes, small mammals and insects or heard the startled gruff 'go back go back' calls of grouse. We rubbed our eyes, blinded by sudden unexpected drifts of smoke and choked on the pungent smell of burning vegetation. Towards dusk we extinguished the last flickers of the fire and walked home wearily, reeking of smoke, with dust in our throats and wood ash in our hair.

I had discovered the fun of teamwork as a dairymaid, but here in the hills, where so much time was spent alone, a special bond flourished between neighbours. It would have been unthinkable to see another herd across the hill and not meet to talk. A gossip over the dyke was an event and a pleasure. Gradually, men I had dismissed at first meeting as old, scruffy, uneducated and poles apart became my friends. Rough tongues and clothes could not conceal for long the qualities and strong individuality of these shepherds.

Inevitably, I became the target for much banter. The herd on the opposite hill to mine carried a telescope and whenever we met, he teased me about the incompetent way I worked my dog. Poor Loch! Within a few weeks I discovered why he had been sold. He was a good dog, working close at hand, keen as mustard and full of eye, crouching, creeping and hypnotising his victims; however, once out of sight, he could not be trusted. When I sent him over the brae to gather a few sheep on his own, my shouts and whistles failed to distract him from his obsessive desire to stalk. His attention became riveted upon one sheep instead of the whole bunch and if in panic she separated and fled downhill to the burn, he was after her without mercy, lunging at her and pulling out tufts of wool. The men said the dog needed a good beating, but Loch could not take punishment. He became shifty and reluctant to work, dragging behind me on the hill and whipping off home when I wasn't looking. On more than one occasion I was reduced to tears. It was impossible to move several hundred sheep up and down this rough, steep country without a dog, or to single off a ewe in need of attention and drive her into one of the round-walled stells which dotted the hirsel. One could run oneself into a state of collapse after these agile Cheviots and Scottish Blackfaces and achieve nothing at all.

Eventually a day came when Loch disappeared and after much searching I discovered him attacking a sheep in the burn. This time he had gone for the throat and ripped not just the wool, but the flesh. The animal was half dead. Hot, tired and beside myself with fury at the dog and pity for the sheep, I took off my leather belt and thrashed him. I wished I could make him understand that I hurt him to save him, for a killer would never be allowed to live. It was no use. A few days later he got away by himself and hunted down another ewe and the manager of the estate said the dog must be shot.

It was a hard beginning. Sheep are such mild, vulnerable creatures with too many enemies. Foxes, crows, ravens, maggots, all lay in wait for the sick and dying. Living eyes were pecked out, flesh eaten away in a seething gouge of blowfly grubs while ticks clung like limpets and bloated themselves on blood. Lambs were abandoned, starved, frozen and buried under snow. Ewes, rolling to scratch their backs and unable to get back on their feet, died ignominiously, four hooves in the air, after hours of struggling. I minded these disasters too much, fretting away the night over a sheep or lamb, which, I felt, I could have saved. But in the morning, when I went downstairs and opened the cottage door, I was consoled.

Sometimes a white sea of mist lay along the valley floor, while the crests of the hills caught the first shafts of sunlight. Often a heron stood motionless by the trout pool, waiting for his breakfast, or a russet fox slipped down through the ferns to drink and, watching his grace, I forgot the lambs he might have taken. Everywhere was so fresh; the air sweetened by moorland plants and the burn water clear and ice-cold. The horse, Kildare, came across the meadow to greet me. He was friendly and inquisitive and inclined to put his head through the window and munch the curtains. And now I had a new shadow to help me instead of Loch.

As a puppy, Tweed was like a plump, fluffy panda, playful and endearing, but that roly-poly exterior did not mask for long all the signs of a fine, working sheepdog. His chest grew broad, giving him the strength and staying power to run all day, his eye was quick and his courage immense.

Long before he was allowed near the sheep, he began to stalk and creep after stray feathers or locks of wool blown by the wind, or lie, nose on paws, eyeing a bird hopping along the grass. Because I lacked a dog, he was encouraged to work as soon as he could walk to the hill and by six months he was indispensable to me. Even the herds who had teased me so unmercifully over Loch, praised Tweed. He'll make you a grand dog, they said, and so he did. I could send him half a mile out of sight and know that quietly and steadily, he would bring back every sheep to be found. Fourteen years later, he was still devotedly penning sheep for me. I could not believe he was dead when I discovered him one morning lying in that characteristic pose, head on paws, but no longer breathing.

A good sheepdog means everything to a shepherd. One of my neighbours, Sandy, who brought a can of milk to my cottage from the farm down the valley every morning, had an old collie called Jed. Sandy was a middle-aged bachelor, short, lean and wiry, with clothes which flapped as though they were several sizes too big. He was a lonely figure and although he liked the company of women, he did not believe he would ever find one who would want to marry him. 'I'm an ugly wee nowt!' he maintained. 'There's no lassie would look at me.' So he made do with Jed, idolising the dog, who was big, shaggy and gentle.

When I first saw Sandy I dismissed him as a tramp, and never gave him a second glance, but as I came to know him better, I realised I had found a friend capable of that rare quality, devotion. Though he might swear his way through every quiet remark and dry, self-disparaging joke, he was a good man and there was nothing he would not do for me. Not only did he bring the milk, but he would chop the kindling, pump the water, bury a dead sheep, or skin me a rabbit. And all the while he worked, he whistled. 'I ken, you're making a right fool of me!' he accused me sometimes.

But then he'd laugh, he didn't care. These were the happiest days of his life, he said.

Sometimes I overslept at the cottage and Sandy and another herd, Jamie, paused on their way to the hill to throw gravel from the burn at my bedroom window and rap on the door with their sticks. Then they'd walk in (for I never bothered to lock the door) and put on the kettle. Jamie was the complete opposite to Sandy: young, dark, good-looking and conceited. He walked with a cowboyish swagger, chewed a wad of tobacco and spat a great deal. Already, at only eighteen, he had fathered a child and was thinking, on and off, about marrying the mother.

He teased Sandy endlessly for not daring to touch me, until eventually, one evening, he goaded the poor man into giving me a peck on the cheek. At that, Jamie roared with laughter. 'Hoots man, that's no a kiss at a'!' Then he demonstrated how it should be done with all the vigour and assurance of a young Casanova. I was not so pleased to be used as a practice ground and sent them both packing.

One day I found the two shepherds on my hill busy digging, while their five dogs looked on. I walked up to see what they were doing and was indignant to discover that they were excavating and destroying a litter of fox cubs. This was my hirsel and they were my foxes. They laughed and, pulling out a second cub, Jamie proceeded to bash it over the head with a spade. I could stand no more. The cubs were about a month old and enchantingly pretty. I made Sandy put the three that were left into a sack and carry them home for me.

That night I shut the cubs in the byre with a bowl of milk and some of the coarse porridge which all the herds fed to their dogs. In the early hours I woke and thought I heard a vixen near the cottage, but I turned over to sleep again, not giving it much thought. When I went to the byre in the morning, I found a vixen had been snooping around and had managed to dig a tunnel right under the foundations to retrieve two of her cubs. The last one scuttled into a corner on seeing me, hunched his back against the wall, bared his teeth and snarled like a demon. The milk and porridge were untouched. When Sandy arrived on the scene, I begged him to try and get a rabbit for my hostile pet and he watched with amusement while, armed with leather gloves, I chased round the byre trying to corner the cub. Eventually, I caught him and held him at arm's length while he writhed and snapped and fought to get free. I decided he must live in the cottage with me if he was to be tamed and not allowed to escape, so I let him loose in the kitchen, carefully shutting the window and door. He rushed about the room hunting frantically for a way out, leaping up on the furniture, scrabbling at the wall and hurling down saucepans. Thinking I was frightening him, I went off round the sheep, leaving the fox cub to explore and settle.

When I returned, I found the kitchen wrecked and sordid, but the cub was still there, crouched snarling under a chair. I handled him again, hoping this would help to overcome his fear of me, but he bit right through the leather of my glove into my finger with his needle-sharp teeth. Sandy produced a dead rabbit and I put out tempting morsels for my wild pet. He spurned them all. He would not eat. It seemed that I had landed myself with a problem child.

After two days the kitchen began to smell rank and foxy, so I blocked up the tunnel which the vixen had made into the byre and put the cub back there. Certainly his mother could not count and having rescued two of her offspring, I hoped she would be satisfied. That night I slept uneasily and in the morning, I rushed out to see if the cub had eaten any food. Cautiously, I opened the door and peered inside, looking for that little, red, sharp-nosed face which spat and swore at me like a wildcat. But the byre was empty. The vixen had come back for him and dug a second tunnel. The

herds thought it was very funny. 'You wait till next lambing,' they said. 'He'll have his own back on you then!'
Secretly, I was glad the cub had escaped, for I had begun to
realise that he was too old to tame easily, his love of freedom
and fear of humans already too strong.

CHAPTER FIVE

Another shepherd friend, Alec, was working on a stick for me, carving the crook head out of a Blackface sheep's horn; shaping and polishing a shank of hazel and fitting the two together with a bracelet of copper. He would allow no one to see what he was doing and hid the unfinished stick away every night under some sacks in the stable. Such care and labour went into this gift. I have it still: the horn head worn thin and the shank renewed, but fit to catch many more lambs and in daily use.

Alec was a shy man who hated gossip. He hoped that no one noticed when he walked or rode his black pony over the hills to visit me, but the other shepherds were well aware of his movements.

If the others did not actually see him, they saw his pony's hoofprints the day after, or his own bootmarks in the soft, gravelly sand by the burn. Jamie and Sandy were like a pair of detectives identifying clues in the tracks to my cottage, the number of dirty cups in my sink, the cigarette stubs in the hearth (for I did not smoke) or the fresh bunches of white heather, which did not grow on my hill. How they loved to tease me! But there was one occasion when they were completely hoodwinked.

It was a Sunday evening and I knew Alec was planning to come over with his fishing rod, hoping to catch enough trout for our supper in the burn near the cottage. Jamie and Sandy had been out shooting rabbits and had called to leave me one on their way home. We were standing in the porch talking when a movement on the skyline of the hill opposite caught my eye. A head and then, the whole figure of a man, stood for a moment silhouetted on the ridge, before dropping rapidly into the bracken. Perhaps I faltered in what I was saying, or stared too hard, but Jamie turned at once to see what had attracted my attention. There was nothing to be seen! Yet somehow I had put him on the alert and now while he talked, he scanned the hill. In vain I threw hints that it was time the men were on their way, but they seemed determined to linger.

'Look! What's that?' Jamie pointed to the hilltop.

'Where?' Sandy saw nothing, but I knew too well at which point to look.

Alec's head was emerging cautiously from the fern to see if the coast was clear.

'It's a fox!' Jamie grabbed his gun and felt in his pocket for a cartridge.

I swung him round violently. 'Oh no! You're not shooting foxes here. That might be my cub.' I glanced quickly up at the skyline; the head had vanished again. 'Anyway, it's gone. Bet it was only a bird.'

I held my breath. After many promptings from me that it was getting late, the herds set off for home. As soon as they were out of sight, the fox stood up on two legs and came hurrying down the hillside carrying a fishing rod!

One evening I made the mistake of inviting two of the herds to supper. As cook and hostess, I had no confidence and little experience, while my guests, who had put on collars and ties for the occasion, were tongue-tied and overcome with shyness. Not until one of them produced a wee dram from his pocket and made us laugh by confessing that his hands were shaking too much to pull the cork, did we begin to thaw out and enjoy ourselves. We were always much happier gathered informally in our working clothes round the cottage fire with our collies strewn about the stone floor. Or if it was summer, gossiping on a sunny, south-facing bank, the men smoking, joking and teasing me with long, feathery-headed grasses. I loved their company; working with them, laughing with them or listening to their endless talk of sheep and dogs.

There was, in Scotland, a riotous side to my life, but it was misleading. I was still naïve and completely unaware that my reputation was becoming notorious and the path to my cottage trodden by too many men. No doubt everyone suspected that someone else was sleeping with me, but I had no time for malicious tongues. Saddened and bewildered by the unexpected marriage of the man I loved, I knew that all I could give and all I wanted was friendship.

Meanwhile, I flung myself into work. I wanted to excel, to be not just a shepherd, but a good one and to match the men in all their skills. At shearing time I was not content to stand and watch between rolling fleeces and catching ewes, but insisted on clipping alongside the herds. A spirit of rivalry flourished. No one wanted to be outdone by a mere girl and as I became practised, so the pace hotted up. Machine shears were not used much in the hills as they clipped the sheep too bare to withstand the elements, so we used old-fashioned shears like big, pointed scissors. It was back-breaking work. The grease in the wool made any cuts or blisters smart, and straightening up after hours bent double was agonising. All the same, shearing was artistically satisfying. Every bite of the shears left its mark on the sheep. It was like making stripes when mowing a lawn. An expert shearer could leave an even, uniform pattern, but the unskilled left tufts and nibbled patches, so that the shorn animal looked moth-eaten and ragged. As each ewe was released, all eyes followed the departing, goatlike creature and cheers and catcalls greeted anything with the chewed-up look, while understated praise like, 'That's a tidy job!' brought smiles of triumph.

Inevitably, a few accidents occurred, a nip here and there, accompanied by urgent shouts for tar. Stockholm or wood tar, a thick black dressing made from pine trees, was used to heal cuts and ward off flies. The wrinkled skin of an old thin ewe or a twitching ear were especially easy to nick, but the worst wound I inflicted was upon myself, when the points of the shears slipped off-course into my thigh. After a week at shearing, I could manage fifty ewes a day, which included gathering the sheep from the hill in the early morning, catching each victim from the pen, and rolling the wool. Nevertheless, I was far from satisfied. I had caught up the slowest herd, but the farm manager was clipping two sheep for every one of mine.

It was always the same: my youth and lack of experience showed me up, whatever the work, and often the men made fun of me, laughing when I picked a ewe which refused to be turned over or kicked the shears out of my hand or escaped with a half-shorn fleece trailing in a woolly train behind her. I did not know whether to be more cross when I was teased, or when I saw the men furtively pushing the smallest ewes in my direction. Yet I enjoyed yard work; dosing, paring hooves, dipping or breeking the hoggets. The last job caused much hilarity. The breeks, sacking squares tied over the tails, were chastity belts to prevent the rams serving the young ewes and the ewes lambing before they were fully grown. Here in the hills, wethers were still castrated by a shepherd slicing off the tip of the scrotum with a knife and drawing out the testicles with his teeth. These were then dropped in a bucket and taken home at the end of the day, along with the fattest lambs' tails, to be cooked for a tasty supper. Sometimes our operator spat his false teeth in the bucket by mistake and the herds suggested that my strong, young teeth would do a much more efficient job. My female weakness prevailed for once; I feared I should be sick and insisted that I was only fit for catching and holding lambs. Surprisingly enough, this old method of castration, performed efficiently, appeared to cause the lambs very little pain compared with the modern ways of constriction with a rubber ring or bloodless cord crushing. Tailing, too, was so quick with the flick of a razor-sharp knife that the only real dangers were fly-strike of the wound later on, or tetanus.

Shepherding is a strange mixture of tremendous physical work alternating with periods of calm, quiet indolence. At lambing time we never thought twice about working all day and half the night, walking blisters on to our feet and bolting out of bed each morning at first light. This was the most interesting and demanding season, which called for all our observation, energy, patience and skill. Anyone who has battled with a stubborn lamb which refuses to suck even when the teat is jammed and squeezed into its mouth, or who has chased round and round a bleak hillside searching for a foolish ewe who has dropped her new lamb without looking behind, will know where the patience comes in. The energy is needed simply to keep going on far too little sleep, for six weeks or more. As for the skill, this is required over and over again, for each individual lambing case is different. Only an experienced hand can distinguish the size, position and aid needed to deliver a lamb, how to disentangle a cord, separate mixed-up twins, or ease out an oversized single. A ewe can stand up to a great deal of manipulation of the unborn lambs within the womb if necessary, but, much more often, help is simply needed to hook forward a folded foreleg in the passage, or to give a last pull on a large lamb when the mother is young, tight, or exhausted.

When I was still inexperienced I found a wild young ewe in trouble with an enormous headfirst lamb. I chased after her; her lamb's head was exposed, swollen and bouncing at every leap. I knew it was alive, for each time she stopped to stare at me, the lamb twitched its ears or its bloated tongue. The only thing in my mind was that I must, must catch her and extract the lamb before it died. Never has a lambing ewe led me such an horrific dance. Tweed could not hold her at bay, and I could not get anywhere near her; she was as wary as a fox and as fleet as a deer. In the end I stalked her slowly, hiding behind boulders, creeping, holding my breath and, at last, I caught her with my crook as the pains of labour compelled her to lie down. The lamb's head was so tightly gripped that there was no hope of hooking out a leg. I could only put a cord behind the ears and pull with a muttered prayer. The sheep kicked and struggled, sweat ran down my back and tears of frustration down my cheeks. I could not shift the monster, and there it was, looking at me with open eyes and tongue turning dark and swollen. There was no one I could shout to for help, not a soul within a mile, just a pair of black crows circling overhead like vultures with some uncanny premonition of the pickings to come.

At first I was afraid of killing the lamb by pulling too hard, but eventually in my despair and thinking that at least I must save the sheep, I used all my strength until I was dragging the ewe herself along the ground. But that did not work either, so in desperation I tried a new tactic. Turning the sheep first on one side then on the other, I pulled in alternating directions and quite suddenly the lamb slid free. A gigantic ram lamb! I shook it by the heels, but it looked dead, which was not surprising after such a prolonged struggle. As a last resort, I blew half-heartedly into its mouth and a few moments later, like a miracle, the lamb gave a bubbly, choky gulp and began to breathe.

In my eagerness to unite mother and son, I made a fatal mistake. I put the lamb on the grass in front of the ewe and released her. She sprang to her feet as though she had never been through any ordeal at all and fled across the hill. This time I had to chase her with her poor neglected offspring jolting in the lambing bag slung over my shoulder. When I had finally got ewe and lamb penned in a lambing shelter within the nearest stell, each had lost all interest in the other and I had to begin at the beginning and teach mother to suckle and son to suck. Natural instinct seemed to have deserted them, the ewe too frightened, the lamb too bruised and shocked. I went away and left them in peace for two hours. On my return a transformation had taken place. The lamb dozed contentedly, its stomach full as a tick, while the ewe licked his damp, woolly coat and muttered with maternal solicitude.

Not all lambing cases ended so happily and sometimes it was necessary to replace a dead lamb with a twin taken from another ewe. The foster lamb had to be dressed in the dead one's skin so as to deceive the mother into thinking it was her own offspring. She recognised her lamb by its smell and if the skin was a poor fit, she circled round and round the impostor suspiciously, or began to bunt it away. It could take as long as a week to break down her hostility. But if the skin fitted like a glove, a maternal ewe could be hoodwinked and the only bafflement came a couple of days later when the lamb's top coat was removed. Then the subtle difference in smell was invariably overruled by the familiar bleat of the lamb, which stimulated the maternal instinct to own and suckle the young one.

The mark of a good lambing is not only to have a high percentage of live lambs, but also to have no ewes in milk running without lambs. Anything with a full udder must rear a lamb. Only the odd barrener can be a passenger. And if I had to define the best asset to a shepherd, apart from a basic and essential love of sheep, I should say that observation mattered most. How much can be achieved by standing still and watching a flock. The ewe with mastitis, the lamb with joint-ill, the aborted gimmer and the abandoned twin all need attention, but milling together in a fold, they could never be picked out from the mob. It takes a practised eye to identify them from their behaviour in the field, or on the open hill. And so when we watched our flocks and herded them into the valleys by morning and out to the high tops by night at a quiet, leisurely pace, we were not really being as lazy and useless as we might have looked.

Our eyes were alert for trouble: disease, strays, mismothering, fly-strike, sheep couped on their backs, or swept into the burns after heavy rain. Our first and foremost responsibility was the welfare of our flock. In the autumn we were expected to take our wethers and surplus ewes into the sale ring ourselves and show them off, hustling them round and round past shrewd and critical buyers. And then we received whatever reward we deserved, for a high price often meant a bonus for the shepherd but a poor bid brought public disgrace. A good lambing, low mortality, a heavy wool crop and high prices at the sheep sales were the final goals. Any stockman worth his salt likes to be as proud of his charges as a mother of her children.

My first season brought me beginner's luck, but my second taught me that no amount of enthusiasm and hard work necessarily ensures success. Sometimes the shepherd fights a losing battle, especially in a hard winter, when snowstorms and prolonged frosts may bring a shortage of food when the ewes are heavily pregnant or the lambs newly born. Unexpected disease, marauding dogs, foxes, gulls and crows may take their toll or a ram may prove infertile and no one knows until the barreners show fat and empty in the spring. But most difficult to accept are the accidental deaths. The ewes which might have been saved if one had looked more carefully, or walked a little further beyond the tump which hid a sheep imprisoned on her back; the lamb suffocated by the membranes left over its face. How I raged at myself sometimes for not seeing, not going far enough, not thinking! And how I hated death! Every shepherd must be a gravedigger, but how sombre an occupation to walk maybe three miles to fetch a pick and spade, and then return to hack a hole out of some stony, inhospitable hillside in order to bury a sheep already ravaged by crows or maggots. Yet I was not sorry to be so involved in birth and death, ease and hardship.

When I had spent two years in the Cheviot Hills, events began to conspire to make me fear that soon I should have to move on. My influence on my friends frightened me. The shepherd Alec was threatening to shoot himself if I did not marry him. Sandy was spending more and more time doing little jobs for me in and around the cottage. Jamie, now married, flirted audaciously whenever his wife turned her back. But the deciding factor was a change in circumstances which curbed my freedom. Accommodation for shepherds was short in the district and it became necessary for a family to move into one half of my cottage. At once the spell was broken; nothing was the same.

I did not return to Scotland for twenty years. Then, on a bleak November night, travelling through the Border country and having nowhere to sleep, I went back to the cottage which I knew stood empty. No one had lived there for twelve years and it was not difficult to get inside, for the windows were broken. How forlorn my old home looked! Slates sliding from the roof, the path grassed over, floorboards rotting, the wallpaper peeling off in ribbons and a musty smell of blue mould. I curled up in my sleeping bag in the driest corner I could find, but how impossible to sleep with memories rushing back and a gale tearing at the slates and through the leafless arms of the ash tree. When I looked out in the morning, it was raining hard, the rain lashed by the strong wind into diagonal lines. The familiar clean-crested skyline and the quick full burn racing past the door remained unchanged, still as beautiful, but the cottage stood like a reproach, unkempt, unloved and too sad.

Yet my return was more than repaid when I went to see Sandy. I found him in his lodgings, still single, now in his sixties, arthritic and bent, but otherwise scarcely altered at all. When he realised who the stranger was, appearing like a ghost from the past, he flung his arms out wide and hugged me. 'Eh, lass, it's grand to see you!' he said and it was as though those twenty intervening years had never been.

He told me all the news. Alec had never married and was still herding on Cheviot. Jamie had umpteen children. The forestry was creeping up into the hills and the shepherds were disappearing. He doubted whether there would ever be another herd living in my cottage. But those two years, so long ago, remained the happiest of his life, he said.

Every Christmas for a quarter of a century, until his death, Sandy sent me a calendar with pictures of the Border hills. 'You shall have one as long as I live,' he promised when I went to see him and thanked him again for remembering me. Such devotion was humbling. For this was the man who had mocked and disparaged himself for being an ugly, ignorant, inferior dwarf. Now, I was the one who felt little.

PART II

ISLANDER

CHAPTER SIX

OPPOSSUM

After I left the Cheviot Hills, a farming friend found me a cottage on Exmoor to rent for five shillings a week. I spent six months there waiting for a passage to New Zealand. The farmer agreed to give Tweed a home while I was abroad and I wanted time for the dog to get to know his new master before I parted with him.

I decided to emigrate for two reasons. The first was a strong desire to see a country renowned both for its natural beauty and for its progressive sheep-farming. The second reason was a despairing resolve to put the greatest possible distance between myself and the man I loved. Now that he was so happily married, there was nothing to keep me in his shadow. I wanted to break free.

It was midsummer, January 1953, when I arrived in New Zealand after six weeks on a crowded emigrant ship. There had been a drought and the hills were yellow with parched grass. The steep, wild coastline looked bare without trees, roads or houses, nothing but dry pasture, rocks and, here and there, a ribbon of sand edging a small deserted bay. This was sheep country, my kind of country.

Wellington appeared suddenly. We rounded a headland, passed a lighthouse and entered an inlet where the hills were clad with trees, and red-and green-roofed houses were scattered among them. We threaded between several small, wooded islets and coming into a shoal of ships, found the city clustered on a steep hillside.

All the immigrants were under contract to complete two years at an agreed type of work, and before disembarking we were shepherded into groups to be issued with instructions and the addresses to which we were to travel. Although by choice I would have headed for a South Island sheep station, there appeared to be no openings for girls in agriculture except as herd-recording officers on dairy farms. Even in this nomadic work, men were more frequently employed and so I found myself standing beside six young men bound for Taranaki.

Jack was one of them. I had noticed him on the voyage. It was impossible not to. Wherever there had been a group of young people talking, laughing, fooling, he was the pivot. He acted in plays, spoke in debates, danced tirelessly, and pursued all the prettiest girls. No one else had so much energy. His pleasing face and open manner won him many friends.

Now he looked at me with a mystified expression, unable to believe that we had travelled on the same ship. Why had he never seen me before? Where had I been hiding?

I smiled at his undisguised astonishment and was both flattered and wary of his eagerness. The ship had been very crowded, I pointed out, and besides, he had been far too busy looking elsewhere.

'Oh,' he said meaningfully, 'you saw me then?'

'You made yourself rather obvious.'

But he took my dig with good humour. 'That's one in the eye for me!' Then after a pause he asked where I was going.

'New Plymouth.' I told him.

'Good,' he said, 'so am I.'

Our conversation ended for the time being, but I suspected he would not let this new-spun thread drop. I moved away amused but indifferent, having observed him often enough to believe that his interest would be superficial and fleeting.

Transport was organised for all immigrants travelling further than Wellington, and those bound for Taranaki were directed to a bus waiting in a yard near the docks. I was one of the first to take a seat and watched through the windows the passengers trickling down the gangway of the ship and looking about them with a mixture of curiosity and bewilderment. They were uncertain whether they liked the appearance of the place, perhaps prejudiced against the tin-roofed wooden houses scattered over the hills surrounding the harbour, and they laughed at the old-fashioned trams in the street. But blue sea, blue sky and bright sunlight influenced us all and on the whole optimism ran high among the prospective settlers.

Jack and another young man called Derek settled immediately behind me and it was impossible not to hear their casual, dissevered conversation.

'I say, look at that by the crossing?' 'Lord! Her skirt's a bit long. Still, they say New Zealand's behind the times.' 'There's a flashy car. American job, isn't it?'

Their chatter never ceased. They were like schoolboys, bounding with excitement. And when the journey began, my attention was divided between looking out of the window and listening to their running commentary on everything of interest which they saw. Most of the time their wonder and criticism and poking fun amused me, but whenever they touched the subject of women my hackles rose, because they talked with carelessness, contempt and underlying laughter, as men do together, showing off like children, pretending to greater experience than they possessed.

The bus climbed noisily and laboriously up the steep road out of Wellington. Over the hill crest the houses gradually thinned and the land spread out before us, undulating and pastoral. It was noon. Jersey cattle gathered round drinking troughs or sat lazily chewing the cud, and flocks of sheep were clustered under trees in pools of shade, panting in their heavy fleeces. A few had been shorn and wandered white and naked in the bright sunshine, nibbling at short turf. The farmhouses lay back from the road looking like suburban bungalows dumped into isolated paddocks, all wooden, one-storeyed, tin-roofed, with fringed or slatted blinds drawn in the windows and empty verandas and neat flowery gardens. No barns or stables, no duck-ponds or orchards of ancient, laden trees. Only milking bails tucked in field corners or big, black woolsheds keeping their distance from the prim painted faces of the houses. And all along the wayside were galvanised mailboxes, with red flags pointing upwards indicating that there were letters to be collected by the local bus. A name was painted on each box: Evans, Apiata, Mackintosh, Willomitzer, suggesting a great mixture of nationalities, and often a placard above the name proclaimed that the *Star* or *Herald* or *Advocate* was read by that particular family.

The main road seemed empty and the side-roads, so often signposted 'No Exit', stretched straight, white and deserted far into the distance. There were no deep, twisting lanes or high hedges. The fields were mostly fenced with wire, but gorse and wild hydrangeas grew along the verge. The grasslands lay open, striped with occasional shelter-belts of firs, and studded with unexpected poplars and weeping willows and native trees unknown to me.

Although sharp-edged hills appeared frequently on the horizon, the road always skirted them and the deep green jungle of the New Zealand bush seldom encroached near the highway. The countryside was more civilised and orderly than I had expected and I looked wistfully at the far hills for signs of a wilder land.

After we had been travelling for a couple of hours, we came to a town where the bus halted and passengers dismounted for refreshments. This was the end of the journey for a few immigrants, including the young man called Derek, and we watched them driven away in cars by strangers who had been waiting to meet them at the bus stop. The rest of us clung together, suddenly valuing the familiar faces which had bored us yesterday. We flocked into a snack bar for sandwiches and tea, or coffee. While waiting at the counter to be served, I found Jack standing beside me, buying cigarettes, chocolate and a rum-flavoured milk-shake. The shopkeeper mixed our change and there was momentary confusion when our hands went out to take the same coin from the counter, then clashed and withdrew, leaving it still there. We turned to each other, both speaking at the same time. 'I'm sorry! Is it yours?'

His eyes were so direct and disturbing, so charged with interest, that I was embarrassed and deliberately sat down at a table with my back to him.

When all the passengers had returned to the bus and the journey was resumed, I noticed a new quietness behind me. Presently there was a crackle of paper and the sound of chocolate being broken into squares. He leaned over the back of the seat with his offering. His hand looked very brown against the mauve and silver wrappings. I accepted a couple of pieces and thanked him.

'What do you think of the place?' he asked.

'It's early yet to judge. I like the space,' I said, 'but so far it's too tame for me.'

61

'Tame? How do you mean? What did you expect, lions and tigers?'

It was difficult to explain that nothing was as raw as I had imagined. The farms were established, the parks and gardens looked mature. I had not yet seen a Maori. Though white people had been in New Zealand little more than a hundred years, they had already changed the face of the land. The virgin country I had hoped to find was gone.

Towards evening, when the sun had set and the light rapidly faded, we came into greener country and saw the cone of Mount Egmont crowning the cream-and-butter plains of Taranaki. The mountain was etched against the sky, a solitary and magnificent pyramid.

'Must climb that one day,' Jack said behind me. It was my own thought.

Only a small band of us were left in the bus, feeling stiff, tired and anxious. The rest had been scattered along the route. Darkness hid the country where we were to live and work, and the last hour of the journey dragged. We peered into the blackness, searching for lighted windows and street lamps, wondering if each small township we entered could be the beginning of New Plymouth. When at last we approached the town, we saw a glow in the sky and knew.

I was the last of the seven herd recorders to arrive at our interview next morning, the men having set off together from the hotel. Jack grinned at me across the room but there was no opportunity to talk. We were swept into a sea of details concerning our work. I had no idea what everything was about, except that we were to travel from farm to farm testing milk for butterfat and recording the yields of the individual cows. I had never seen a centrifuge or heard of the Gerber test. Yet within a couple of hours, I was given a district, set up with a vanload of equipment and pronounced ready for the road. An instructor was to accompany me for the first few days.

We left the town and drove along a straight road across pancake-flat country. We passed a couple of small butter factories and a meat-canning works, which filled the air with a horrible stench of cooking flesh. My spirits were at a low ebb. There were dairy farms everywhere, each with its bungalow set in a paddock. Several times I noticed Jersey bull calves, only a day or two old, standing tucked up and miserable in crates or pens, waiting for the cattle lorry to take them to the slaughterhouse. The female calves were used as replacements in the milking herds, but for the majority of the bulls there was no reprieve.

A Maori boy rode past us on a horse with a couple of dogs running behind. He waved a hand. So did a man driving a bulldozer. People had time to greet each other and traffic was so thin that heads always turned to stare at a passing vehicle. We came to a rise and twist in the road and suddenly, unexpectedly, caught a glimpse of the sea – a blue triangle between hill shoulders. I had not realised that we were so near the coast, but in fact our route ran almost parallel to the great, curving bay of New Plymouth. Far ahead of us a long bank of grape-blue cloud gradually defined itself as a distant range of hills. 'The King Country,' my companion said. A wilderness of jagged, sharp-spined ridges, darkly clothed with thick, impenetrable bush.

Presently we turned down a side-road seawards and I learnt that this marked the beginning of my area of farms. Here I should be living on the very edge of the dairy plain, where it merged into the foothills of sheep and timber land, and close to the sea, where the strange black sands stretched for mile upon mile, unvisited, untrodden. No beach huts, no deck-chairs and ice-cream vendors. Nothing but seagulls and surf and perhaps a far-away fisherman waiting, like a cormorant on a rock.

The farm where we were to work that afternoon and stay the night was right on the coast. When we arrived at the milking bail and got out to unload our gear, there was a wonderful seaweedy tang to the wind. We set up a series of buckets in a line down the cowshed, attaching each to a milking-machine unit, and laid out the record books, sample bottle, pipette and scales in the dairy.

Later we walked across a paddock to the farmhouse for tea, where I met the farmer's wife and three school-age children. They were friendly and innocently curious. How did I like New Zealand? Had I ever seen the Queen? Did I know their second cousin who lived in Bristol? Did I play bowls? Apparently bowls was a favourite with young and old. And what did I think of the All Blacks? I had never heard of them, but evaded saying so, sensing their supreme importance to these people.

A barking of dogs brought us to our feet and outside we saw the cattle, in a cloud of dust, coming along a track to the milking bail. A boy who helped the farmer was walking in the rear, barefoot. The cows entered the shed ten at a time and as each animal was milked, my instructor showed me how to weigh, sample and record its yield. After the morning milking, we would test the samples for butterfat content. Then, having written up the records, washed and stowed our equipment, we would lunch and move on to the next farm. There were twenty-six farms to visit each month, and at the end of the round, four or five days of freedom. It was to be a suitcase life, a different bed every night, but no roof promised when on leave. I wondered where I should go and what I should do when that time came.

Not all the farms proved as agreeable and homely as the first, and after I had been launched on my own, I visited one which appalled me. It was a poor place, set in a deep shaded valley, where bush encroached down the hillsides into the damp rushy pastures. The small herd looked thin and produced low yields, and the dairy and cowshed were neglected and badly equipped. A smell of rats pervaded the building and I had to light an old boiler to heat water for washing my equipment. My van got stuck in the muddy yard and in the early morning, when it was still dark, I had to write the records by the light of a candle which was standing in an old cocoa tin. The house was an unkempt wooden shack with an overgrown garden littered with broken toys, old bottles, tins and cigarette stubs. There were five young children and I shared a bed with two of them. The fleas were unbelievable and I scratched my way through a sleepless night. At mealtimes the children stuck their fingers into everything and the farmer carved the bread clutched against the sweaty singlet in which he had milked the cows. Nobody offered me water for washing and it was only by hunting about the place that I discovered a doorless hut at the bottom of the garden, containing a foul pit, a swarm of flies and a pile of mildewed newspapers. There was only one compensation: a chance to explore the native bush for the first time, to see the wild pigs and possums, the strange trees and ferns and jungle of creeper, and hear the wonderful variety of birds. Everything was so new and exciting that when I managed to escape from the farm in the evening for a walk, I was entirely happy and absorbed.

At the end of my monthly round of farms I took advantage of the few free days to explore further afield. Since my van was provided for official work only, I had to travel as best I could, walking, hitch-hiking and resorting to buses and trains when other means failed. On one of these random journeys northwards over the convoluted ranges of the King Country, I took a lift with a sheep-farmer who told me about a remote island off the Coromandel Peninsula which he was planning to buy. His description of the place fired my imagination and spurred me to head directly for that far coast. It proved a marathon trek but I was determined to see the island.

All traffic faded away as I neared my objective and the last eleven miles were covered on foot. I was so tired I dared not pause to rest. It was almost dark. Wild pigs scuffled in the undergrowth and moreporks or New Zealand owls began to hoot. After a weary uphill climb, I came to the crest of a high ridge and looked down on the sea. Beyond the bush-clad slopes and rocky indented coastline, the vast expanse of water stretched to meet the sky. Away to the east a light flashed. I waited, tense with excitement, and presently it flashed again. Straining my eyes in that direction, I could just make out a shape surrounded by sea: a solitary black slug of land. Footsore but jubilant I followed the coastal road down to a small township where I found a bed for the night.

Arriving at dusk, I had no clear picture of the place until daylight returned. Then I saw that it was beautifully situated beside an estuary which widened out to resemble a lagoon. Rugged, forested hills gathered round the water, mangroves grew in the shallows and, on the ocean side, the river cut a narrow outlet to the sea through a mile-long bar of white sand-dunes.

I walked down to the jetty and found an old man tinkering on his fishing boat. Though the tide was too low to cross the bar, he offered to take me to the estuary where I could look at the island through his binoculars. Soon we were chugging down the river, sharing a bag of passion fruit and trailing a fishing line with a spinner in our wake.

Approaching the bar, the view changed, revealing the breadth of sea beyond the low sand-dunes. And then I saw the island. A solitary wedge of land. The old man pointed to it and handed me his binoculars. It was further away than I expected, and hazy, but this made it all the more tantalising. I could see white surf breaking against the rocky northern point and, in contrast, calm water in a scalloped bay edged with sand. A ketch was anchored there and the house was pinpointed, sheltering among trees halfway up a hill, by the sunlight glinting on the glass of the windows. A lighthouse crowned the summit of the island.

Immediately I began to dream of living there.

We anchored in the lee of the dunes and fished for an hour. The fish were abundant and the water was so clear that we could see the shoals of snapper swimming round our lines. The fishy salty smell and the raucous gulls swooping overhead reminded me of childhood holidays by the sea.

Before we returned up the estuary, I looked back at the island, memorising its triangular shape, the single ship in the bay, the trees dark and shadowy against the paler pasture, and the high needle of the lighthouse. A seed was sown.

Two days later I walked down the high street of New Plymouth. I was going to the bank and making a few purchases before returning to my round of farms, when I heard a voice behind me.

'Thought I recognised your hair. I ought to . . . I sat behind it all the way from Wellington.' It was Jack.

We were standing on the pavement in midstream, people swerving to left and right. He caught my arm and drew me to one side. 'Look, we can't talk here. Let's go and have a coffee somewhere.'

So we found a small milk bar and I sat down to wait while Jack fetched food and drinks. The room was stuffy and full of flies. Crumbs were scattered on the red-topped table and someone had spilt their tea. Plates of tired rolls and lurid cakes under glass covers lined the counter beside vast jars of brilliantly coloured cordials for flavouring milkshakes.

Watching Jack waiting to be served, he seemed younger than I remembered and unmistakably English. He was wearing the kind of tweed sports jacket never seen on a New Zealander. Harris or Cheviot probably. Feeling my eyes, he glanced round and smiled. There was no doubt about it, he was worth looking at and this added considerably to the pleasure of being in his company. I could see the girl behind the counter assuming the special coy, yet come hitherish expression which she reserved for presentable young men.

He came back to the table with two coffees and a plate of biscuits, and sat down opposite me. 'Sorry,' he apologised. 'It's rather grim, isn't it?' Then he steered the sugar bowl round the spilt tea towards me and went on eagerly, 'Well? Tell me everything. What have you been doing? Where are you working? How do you like this herd-testing racket?'

I told him that I was lucky and had a wonderful district on the edge of the King Country between the hills and the sea. I described how I had spent my leave hitch-hiking but I did not mention the island. Dreams had a strange way of collapsing if advertised.

We compared notes on New Zealand and laughed together over the ravenous fleas, the lack of locks on lavatory doors, the milking to music and those predictable questions: had we seen the Queen? Did we know some distant cousin in Manchester? And then the food: the eternal mutton, silver beet and pumpkin, followed by bottled peaches. But whereas I was happy and excited with my life, Jack was restless with discontent. Already he was bored with his work. He wanted something more creative, mind-stretching, challenging. He had even been allotted a horse and cart to carry his herd-testing equipment, he said, instead of a van.

'I wouldn't mind swopping. I like horses,' I told him.

'Wish I was a horse!' he sighed. Then he asked if I was going to the Herd-Recorders' Dance.

I said I hadn't thought about it.

68

'Think now.' His eyes, direct and eloquent, beseeched without restraint.

I explained that I was not much good at dancing, but eventually his persuasion won. Triumph was written across his face: an appealing face, expressive and incapable of concealing anything . . .

I returned to my round of dairy farms. How tranquil it was in the early mornings watching the cows coming in over the flaʈ, green meadows, smelling their warm breath as they queued on the doused concrete approach to the milking bail. I worked quietly, filling my sample bottles, ticking off names or numbers and in the pauses between cows, watching the farmer washing udders and attaching machines. There were always a few farm cats waiting around for a saucer of milk, mewing with hunger and then sitting licking their paws and wiping their faces, while outside the cowshed a cattle-dog would hunt up the last slow members of the herd, barking round their heels and urging them into the bail. In spite of the effort of early rising, I liked the morning milking best. There was a freshness about the farms after a night's rain or heavy dew, and when the work was done, we went in to breakfast, hungrily appreciative of the food and hot tea awaiting us.

Not until the midday interlude, between leaving one farm and arriving at the next, did I have leisure to think of other things. The island obsessed me.

A few weeks later Jack persuaded me to go to a cinema with him.

The film was appalling, so bad that at least we could laugh and poke fun at it. We came out of the building, the smell of stale smoke clinging to our clothes. It was raining hard. The small township offered no other entertainment, so we bolted through the downpour to Jack's car. He had bought a midget Austin. It had cost £30, and I could see why. Rain had got into the engine, so that it refused to start, and had leaked through the roof, filling the seats with pools of water. We mopped them with crumpled sheets of newspaper and got inside. The lamplit street was deserted, the gutter a running stream, and the wind-screen rapidly steamed up. We laughed over the film and we laughed at the rain dripping on our heads. Then suddenly, out of the blue, Jack asked me to marry him. I was dumbfounded. How could anyone in their right mind want me as a wife? A more reluctant housekeeper and a lousier cook would be hard to find. Didn't he know that I lived in the clouds with my feet hardly touching the ground at all, that my head was full of far more exciting plots and plans than marriage? Did he really think that I could be content with a duster and dish-mop in either hand and a baby squawking in a pram? The moment had come to explain as gently as I could that I was in love with a married man in England, and that I was obstinately heading for my own particular sort of nunnery – an island! An island and a flock of sheep.

Jack took this announcement very well. If he had possessed any money, he would have bought me an island there and then, he said. But in truth he had nothing in the bank, no house, no savings, nothing but a car which was neither watertight nor mobile. He could see that his proposal of marriage was as preposterous as my own ambitions and that I did not even believe that he loved me. Well, time would prove which of us was in earnest.

During the next month I meandered through my work, brooding about islands and wondering if Jack really meant what he had said. Then, one day while I was unwrapping a newspaper from around a new packet of soap powder, I glanced idly over the pages, realising that I had almost lost touch with the outside world. Suddenly three words caught my attention: 'Island for Sale'.

70

A farm by the name of Aroa had been put on the market. An island property of one thousand acres, suitable for sheep. I fetched out my map of New Zealand and found it: a dot in the far north, miles from anywhere. It was certain to be hopelessly beyond my financial reach. And it was inconceivable that this island could be as perfect as the Coromandel one. Nevertheless I was out of my mind with excitement and already plotting that magnetic course due north. I had bought a second-hand van and at Christmas I should have nearly a week of liberty. I knew exactly where I was going.

CHAPTER SEVEN

O n the third day after Christmas I drove seawards along a dusty contorted road. The country was dry and neglected, traffic almost non-existent, houses few and the farms appeared rough and impoverished. Maoris outnumbered whites and they were much darker than those I had seen in Taranaki. Their way of life looked more primitive, their possessions limited to a wooden shack, a patch of maize and sweet potatoes, and a lean horse or two tethered on the verge. It was a different landscape from the rich fertile dairy plains. Acres were lost to manuka, a scrubby native shrub which, unchecked, spread voraciously. Here in the north the heat was fierce and I longed for shade, water and rest. The road was like a choppy sea; the van bounced and jolted all the way while everything was coated grey with dust.

In the early afternoon, feeling weary and pessimistic, I came to the crest of a high hill overlooking the coast. And then, between the powdered fronds of punga trees growing by the dusty wayside, I caught a first glimpse of Aroa. The island rose from the sea in a golden dome with long ragged reefs radiating outwards like the spokes of a wheel. Ribbons of white sand edged the bays and shadows patterned the gullies. A scattering of islets, rocky and steep, encircled the main island, which in the hazy air lay more like a cloud than land across the water.

I braked reluctantly to descend the steep hill and presently came to a heavy gate barring the roadway, which by now had deteriorated into a rough sandy track. An old man slipped silently from the shade of a group of bluegums and came forward to open the gate. He must have been standing there watching the vehicle approach. His dark weathered face suggested Maori blood, but he was dressed in new white tennis shoes, shorts and a large straw hat, and had a strange air of suburban respectability. I learned later that he was the chief of the tiny fishing settlement.

I asked him if the island was still for sale. He pondered. His speech was slow, his quiet words drawn out like sighs.

'No, it was sold three days ago,' he said. My heart slumped. 'You want an island?'

I nodded, but without hope.

His mouth split into a toothy smile, 'I can sell you an island.'

'Where? How big?'

He waved an arm pointing across the sweet potato ground, leafy maize patches and white undulating sand-hills, to the sea beyond and a small rocky island.

'But there's no grass,' I protested.

'Plenty of turkeys,' he said as though these would compensate for everything.

I shook my head. 'I want to farm,' I explained. 'I want to have sheep and cattle.'

'Sheep and cattle!' He held out his hands palm upwards and then let them drop languidly in a gesture of despair and

74

mystification. 'You want too much,' he said simply, then raised his hat and stepped aside for me to pass.

The road was too narrow for me to turn the van, so I had no alternative but to continue a further half-mile, where it sidled over a rise and ended at a small, shingle beach. After driving most of the day through dry, deserted scrub country, I drew up in surprise to find myself on the verge of a crowd. At least a dozen Maori children were playing in the shallows, squatting on the pebbles and clambering over the rocks. Beside a rambling line of wooden shacks facing the shore, dark men and women sat chattering and smoking. A group of fishermen were leaning over their boats, gutting their catch, while gulls screamed hungrily overhead and three mongrel horses harnessed to sledges stood patiently in the hot sun, swinging their tails at the flies.

The air reeked of fish. There were nets drooped over poles and the brown glass of empty beer bottles glinted among the litter. The tin roofs of the houses beat out the heat and where paint existed, cracked and swollen blisters spread like a rash over smooth skin.

Across three miles of calm water lay Aroa, aloof and golden. Isolation gave it a veiled enchantment. It was even more beautiful than the Coromandel island: higher, wilder, like a star-shaped mountain with the points running down as reefs into the sea. Trees grew in the gullies and the white dot of a house was just discernible close to the shore, in a sheltered bay. The tiny islands clung around Aroa like the beads of a necklace; the beds and breeding grounds for seabirds.

It was no use staring wistfully. The island was sold. My attention was drawn to a large man sitting in the middle of a cluster of Maoris. His skin was rosy, wind-thrashed, and his jaw and chin were hidden by a thick sandy beard. He was an Irishman and was the late owner of the island.

'Why did you sell out?' I asked curiously.

75

'I've always been a wanderer,' he explained.

He was a man of about fifty. He looked well fed and well pleased with himself. There was good reason for his satisfaction. Having sold the island, he had bought a ship and was shortly to set sail for Ireland. He told me that the purchaser of Aroa was a professional man from the south who intended to use the place for fishing holidays.

'Won't he be farming the land?'

The Irishman shrugged his shoulders. 'Maybe he'll put a man over there. I couldn't tell you his plans.'

I thanked him and returned to the van. Before I could forget, I wrote down the name and address of the island's new owner. If no one was going to be farming, or living permanently at Aroa, there might be a chance of leasing the place.

Now I was all impatience to send off a letter. I took a last, long, covetous look at my goal rising steeply from the brilliant water. A dare, a dream and a challenge. I could have hunted the whole world over and never in a lifetime found anywhere so right: warm, high, pastoral and severed by the sea.

I wrote at once to the owner asking if he would lease the property to me or, failing that, consider a farm manager. I stated my age and agricultural experience, but did not reveal my sex. Having posted my letter, there was nothing more I could do, but hang around the local post office like a cat beside a mousehole.

A week later I found a letter in strange handwriting waiting at the post office. I tore open the envelope as I walked out of the building. The contents were brief and to the point. The owner of Aroa would not lease the farm, but he was prepared to consider the possibilities of share-farming, if I would send him references first.

I warned myself again and again that as soon as I announced my sex, I risked the collapse of the scheme. Nevertheless, I posted references and this time signed not just my first initial and surname, but my whole name. Then with inner prayers and fingers crossed, I waited for a reply.

Early the following evening a telegram arrived for me. It read, 'Still interested. Can you come for an interview?'

At the end of the week I drove south to Palmerston where the owner of the island, a doctor, lived and worked. His house was a typical New Zealand bungalow with red corrugated-iron roof, cream-painted wooden walls and a large well-tended garden. He opened the door himself and showed me along a passage into a sitting room. I was relieved to find a man old enough to be my father. His face was deeply lined, his grey hair thinning and he wore spectacles. He invited me to sit down, looked at me with a pair of restless eyes and then, rolling himself a cigarette, he paced about the room, waving his hands and talking without a pause. I was hardly called upon to utter a sound.

He was full of idealistic dreams for the island. He planned to plant thousands of trees, make a bird sanctuary, dam a stream to form a lake. He hoped to build a new house on the top of the hill, where the view was magnificent. All ploughable pasture was to be reseeded and he would introduce pheasants and breed prize-winning cattle. He admitted he knew little of farming and when it came to discussing finance, he made it plain that he was not a wealthy man. In fact the purchase of Aroa had emptied his bank balance and there was no question of him giving up his medical work for many years. Indeed, he doubted if he would ever live on the island permanently, as it was too isolated to suit his wife.

He was adamant that if I went to the island, I was to accomplish all that a man would. If not, I was to pay for men to help me. He also let slip that he had tried in vain to find a manager for the property. He could not conceal his astonishment that I should want to go there. And alone!

That was the part which worried him. Supposing I was taken ill, or had an accident? How could I possibly cope with some of the jobs like rounding up stock, handling heavy fenceposts, winching the boat up the beach? One moment he was telling me that I could not possibly manage and the next that I must. He eyed me dubiously, bothered by my youth and dress.

Eventually we struck a bargain. In return for grazing up to 1,000 sheep of my own, including not more than 500 ewes, I would look after the island, fences, boat, machinery, house, gardens and the 120 cattle which the Doctor would be taking over from the previous owner. He would pay me £3 a week, sufficient to live on, until I received my first wool cheque. Neither of us really knew in whose favour the scales tipped. I would have agreed to anything in order to live at Aroa and, for the moment, it was irrelevant that I had far too little money to purchase as many sheep as I should be permitted to keep. My only concern was to set foot on the island.

Possession of Aroa was to be granted on 2 April, though the Irishman with a roguish sense of humour had offered to vacate the property on the first. The Doctor and his wife planned to spend a weekend at Aroa in a fortnight's time. I was invited to accompany them, in order to taste island life briefly, before making my decision final. This was a generous offer which I accepted, though privately I did not believe that anything on earth could make me change my mind.

The Doctor and his wife made a detour into Taranaki to collect me on their journey to the island. They arrived one evening and we drove all night, all next day and most of the following night. Our progress was slowed down because, on a trailer behind us, we towed a boat containing crates of pheasants. There was also a mountainous pile of luggage, crowned by a spaniel on the back seat, so that the three of us

78

were compelled to sit crushed together in the front. It was now the tail end of summer, but still uncomfortably hot in a car at midday and we became very tired and stiff. We lived on a monotonous succession of sandwiches, eaten as we went along, with occasional cups of tea brewed by the wayside. The Doctor would stop for nothing except urgent calls of nature, or to tend his pheasants or to stamp up and down the verge to rid himself of cramp. After such intervals, he returned to the driving seat temporarily refreshed and entertained us with many renderings of 'The road to the Isles'. And sometimes we were laughing, sometimes we were arguing and sometimes there was a long weary silence. Once he asked me to give him a rest from driving, but my performance did not please him. He needed three whiskies, he said, to be a passenger with me at the wheel. And so he drove and drove until he could hardly see out of his eyes, or lever his body from the seat without groans of agony.

His wife was a kindly, gentle woman. Many years of marriage seemed to have taught her to endure her husband's eccentricities without retaliation. When she asked tentatively once or twice if we might not stop for a proper meal and stay at an hotel on the second night, his stubborn refusals soon silenced her. His temper was quick to ignite and he got flustered and irritable too easily. Her role was to soothe him, but towards the end of this testing journey, her large vulnerable brown eyes blinked back tears as she grew more and more exhausted.

We arrived at the tiny fishing settlement opposite Aroa in the early hours of the morning. It was raining and windy. However, there was a shed at the top of the beach used as a store and garage by the island's previous owner and here the Doctor's wife and I bedded down between the petrol cans and fishing nets. But not the Boss. He pottered about all night, feeding his pheasants, grieving over the five which had died on the long journey, unpacking luggage and examining the rough sea by torchlight. Beyond his movements came the sound of the sea: the wonderful soughing, sucking and booming of a relay of waves smashing on the sand, pebbles and rocks. And in the air hung a reek of fish, wet stones, kelp and kerosene from the tins at my feet.

When I looked out of the shed in the morning, the Boss was toiling up the beach with a load of driftwood for the fire, his canvas shoes flapping and a pheasant's feather jammed in his shapeless hat.

'Hell's teeth!' he said to me in greeting. 'I don't like the look of the sea.'

It was a grey and white sea, the waves steep and short. A gusty wind blew against the strong current flowing between the mainland and the island. The water churned angrily over the reefs and spray rose high above the rocks on the windward shores of Aroa and the smaller isles. Within the comparative shelter of the bay before us the breakers rolled in less ominously, but there were no signs of any fishermen launching their boats.

Eventually, in his own good time, which was certainly not early, the Maori for whom we were waiting made his début. He padded up silently and looked in through the open shed door where we were gathered in consultation.

'Good morning! Good morning!' He cried in a breezy, jovial voice. 'She's boggy today.'

We realised he meant the sea and leaning forward the Doctor asked anxiously, 'Will we get across?'

'By Joves, that won't stop us!'

And we all sat back with relief. Ben was the most reassuring figure. He was sturdy and corpulent. With sixty years of experience behind him, he exuded geniality and confidence.

I had been unable to repress feelings of alarm at the idea of my neighbours being Maori, so dark and alien, but now confronted with Ben, such thoughts were laughable. His whole personality was so cheerful, so heart-warming. One liked him instantly. Physically he was stalwart with thick muscular arms, large feet impervious to the sharpest stones, and a broad chest. He wore shorts slung dangerously low round his huge belly, a ragged shirt and a singlet. His eyes brightened at any mention of alcohol and his world was taken up with fish, boats and the sea. Besides possessing the biggest boat, he owned several horses, a cow, and a patch of land for growing maize and kumaras. As a seaman and the head of a large family, he was an important person in the Maori settlement.

The Doctor suggested that Ben should take me and the pheasants to the island first. His own boat was much smaller and flimsier than Ben's sixteen-footer and he wanted to postpone making the crossing for an hour or two in the hope that the wind would drop and the sea quieten.

Once out of the shelter of the bay, the voyage became exhilarating. We rose and plunged over the waves, spray in our faces, the bows shooting up in the air, before diving into the next trough. I could taste the salt in my mouth and feel my hair dripping into my collar. Sometimes we sliced off the curling crest of a wave and water cascaded inboard, but Ben was whistling, unperturbed, and I was too happy to be nervous, my attention riveted upon our destination.

Aroa loomed bigger and wilder than I had expected. As we drew nearer, it became possible to see details which were not visible from the mainland. There were outcrops of rocks and the red-brown scars of landslides on the shins of the hills. The grass was tussocky and encroached upon by reed and manuka. Much of the shore was strewn with boulders or shadowed by ominous cliffs where the sea had gouged dark caverns out of the stone. And patches of frothing spume whitened the water over submerged reefs.

Then, contrasting with these almost frightening aspects of the island, there appeared a wide half-moon bay, so sheltered by curving headlands that the water was only ruffled by the wind. Sands, rippled by the ebbing tide, stretched in a golden crescent and along the rim of the shore grew deep-green pohutukawas and strange shaggy cabbage trees. The gullies led inland and among the scrub and flax bushes growing near the streams grazed red and white Hereford cattle. They lifted their heads and stared at us with pale faces as we chugged into the quiet lagoon and drove the bows of the boat on to the sand.

Here we unloaded the crates of pheasants and carried them up the beach, through the jumble of driftwood which lay at high-water mark, to the grass. When we released the birds, the cocks flew up with a whirr of wings and headed into the gully, but the hens scuttled into the undergrowth of rushes and ti-tree. Ben eyed them appreciatively, thinking no doubt that they would make a good supper. I looked all around me, longing to explore; dying to see over the shoulder of the hill.

Presently we returned to the boat and, leaving the bay, rounded a rocky point, where a group of shags stood like black sentinels, and came to another bay, less beautiful, greyer, almost austere. The beach was steep and stony and strewn with occasional boulders. And then I saw the homestead tucked in the lee of a fir-clad hill. Apparently the anchorage was safer there and the shelving pebble beach made it possible to approach close to the shore, whereas, at low tide in the sandy bay, the shallow water receded a hundred yards, hampering the launching or winching in of a boat.

The farmhouse was a one-storey white wooden building with a red tin roof, standing within its own shadow's length of the sea, so near the water that a fishing line could be cast from the window. Shingle had been heaped in undulating mounds against the seaward wall, but on the other three sides of the house lay a garden and there were bamboos and peach and fig trees growing in a windbreak on the boundary. Further along the shore stood a boathouse and a large woolshed where the sheep were shorn and the wool packed. Nearby were the dip and stockyards, enclosed by the kind of strong, high wooden rails seen round a ranch corral. Behind the farm buildings I saw a small paddock where a black cow and four horses grazed. A rough track, cut into the steep hillside, led inland towards the core of the island.

Later in the day, after the Doctor had arrived in his boat, we borrowed a couple of horses and rode round the property. The island was just over two miles long and one mile wide. The horses picked their way along narrow sheep tracks where the land climbed steeply on one side and on the other fell away into rocks or water. In places, the cliffs were precipitous, dropping down hundreds of feet to the booming, echoing surf. Elsewhere, there were bays, half-moons of tide-washed sand, tantalising and inaccessible. On the northern ocean side of Aroa, we came to a magnificent beach with white sand-dunes and, after tethering the horses to a flax bush, we walked along the shore. Gannets dived like white arrows into the water and when we reached an outcrop of rocks, we saw a colony of oysters embedded there. I had not yet acquired a taste for them, but the Doctor whipped out his sheath-knife and began to prise them open and then devoured the salty gelatinous contents.

We returned across the centre of the island, galloping along a spine of hill, then skirting a large area of bush which clothed the steep valley head. Somewhere within the trees a stream sprang and we could hear the sound of a waterfall. Here lived hundreds of small birds, tame and sweet-voiced, a different race from the big fish-eating seabirds. Although the grass had yellowed in the summer drought, there were still many green clearings in the moist and shaded gullies and the cattle, drawn to these places, looked fit and content with a bloom on their coats. We found a few banana trees planted in a sheltered sunny combe and grapevines bearing shrivelled forgotten clusters of fruit.

That night the wind dropped and a brilliant moon shone over the sea. I was lulled to sleep by the quiet, rhythmical stirring of water and pebbles within a stone's throw of the house. And the following day only confirmed my certainty that this was the place for me.

When morning came, the sun was shining as it had not been on our arrival and the sea had calmed. I went out early and swam at Waiiti, the sandy bay where Ben and I had released the pheasants. Shoals of slim, pointed fish were leaping through the shallows chased by bigger fish and Little Blue Penguins, and a flock of gulls floated like a white cloud on the water. The sea was a tropical blue-green; diamond clear, revealing every dark weed and pale shell. And the island itself was infinitely peaceful, the hilltops catching the mounting sunlight, the gullies still in shadow. Some of the cattle had come down to the beach to pick over the bronze kelp abandoned by the ebbing tide and they stood there chewing the cud and watching me. The Doctor was watching me too, as he lay on the grass at the top of the beach, smoking and beaming, but I was too happy to care. He had come to look for his pheasants, he said.

He gave me my first lesson in handling a boat that morning, but his initial demonstration of how to start an outboard motor was spoiled by his forgetting to switch on the fuel. And a little later when he set off with the anchor still in the water, my inferiority complex concerning everything nautical gained a measure of confidence. Nor was it quite so alarming to be entrusted with the tiller of the dinghy that afternoon to convey the Doctor and his wife and the Irishman's wife to the largest of the small islands encircling Aroa.

Shag Island was a treeless wedge of land set in a rock-strewn sea. High, sheer cliffs dropped into thick, inky water on one side and on the other fell away in a steep slope to a

84

rocky shore. There was no shelter anywhere and only a short, straight ribbon of beach where it was possible to land in calm weather. When we stepped ashore, we trod among shells, powdered and whole, all colours, shapes and sizes. There were tiny red fans, black-and-white ringed barrels, purple cones, silvery grey saucers and the lovely fragile green cases of sea urchins. The Irishman's wife had come to gather a few last shells to take away with her and while she and the Doctor's wife combed the shore, the Boss and I climbed to the top of the island and looked down the precipitous cliffs to the shadowy water below. Then, turning thankfully away, we paused to regain our breath and sat in the yellow, tussocky grass, facing the sun. The long, undulating line of the mainland ranges lay before us, a patchwork of farmed and virginal land, and away on our right Aroa rose from the sea, with headlands and reefs running out like spokes into the water.

'Well?' said the Doctor. 'What have you decided?'

I replied without hesitation. 'I want to live here.'

Events moved fast now. We returned to the mainland the next day and I began to search for the sheep which would form the nucleus of my flock. The major sheep sales of the year had already taken place and it was not easy to find anyone who wanted to dispose of breeding stock. Very soon the gossip spread that I intended living on the island and wanted to buy sheep, and on the second day a strange man stopped me in the street to tell me he knew of a place, twenty miles south in the hills, where a farmer had a couple of hundred young ewes to sell. I went down to see them and bought them that afternoon, having haggled for half an hour, leaning against the yard rails, looking down at their jostling woolly backs. People said I had a bargain, but I was not so sure. I was too well aware that I had spent every pound I possessed and a few more, borrowed from the bank.

Before leaving the far north the Doctor and I decided to sign an agreement and accordingly searched the town for a suitable solicitor. Eventually we found ourselves in a small, gloomy office, peering over imposing tomes at a startled lawyer. Never in his life, he told us, had he been asked to draw up the kind of contract which we wanted. But after half an hour's thought and consultation between the three of us, the unique document was composed and the draft copy signed and witnessed. Triumph overrode an inner whispering that I had burnt my boats irrevocably. We returned to the hotel in high spirits to drink toasts to our bargain, to each other, to the sheep and to Aroa. While the Boss lamented the fact that he would be too busy to spend more than a few weeks on the island during the whole year, I secretly rejoiced and celebrated his eccentricity. Probably nobody in their right mind would have taken on such a youthful daydreamer. Glowing from a combination of alcohol and elation, I hugged the thought that the island was as good as mine.

CHAPTER EIGHT

After completing my last round of dairy farms in Taranaki, I packed my bags, having arranged to rejoin the Doctor at Aroa on 2 April. On the road north my thoughts were all for the future. Jack had decided to accompany me until I left the mainland. His curiosity had got the better of my discretion. He must see the island, if only in the distance. My plans worried him, but he was keeping himself on a tight rein, no longer trying to influence me, and I assumed he was resigned to my going away.

A Border collie called Glen, a gift from a local farmer, and Jock, a Huntaway puppy Jack had given me, sat in the back of the van among my cases. We took it in turns to drive and tried to doze when not at the wheel for it was a long, slow journey. I was keyed up with fright and excitement. Soon, unbelievably soon, I would be living alone on an island.

We arrived in the late afternoon. As we reached the crest of the hill overlooking the coast and the small fishing settlement, I pointed eagerly: 'Look! Look! That's Aroa.'

The sky had clouded over and the sea was grey. The island appeared further away than I remembered, aloof and beautiful, but just a little more intimidating to me than I would have liked to admit.

'You can't live there alone,' Jack said, shocked.

'I'm going to.'

The gate stood open, so we drove past the Maori chief's shack and the tin church nearby, then cautiously over the sand-hills. When we reached the shore, we found the Doctor and his wife in the car shed. They had arrived earlier that day and were waiting to take possession of their property next morning. The Boss advanced to greet me. On seeing Jack, he was momentarily disconcerted, but quickly rallied.

'Come and have a drink,' he said and, leading us into the garage, drew up a couple of fish boxes for us to sit on. Then he poured out four whiskies and handed them round.

I forget how many refills had been consumed when the men began to sing. Jack gave a rendering of 'The road to the Isles' in a Scottish accent at the Doctor's request. He could sing and joke without restraint and I could tell he was making a hit.

Later the Doctor said to Jack. 'You must come over to the island with us tomorrow and have a look round before you take off for the south.' And then bringing out the bottle and looking from Jack to me, 'Would I be presumptuous in suggesting we drink a toast to your future?'

I glowered at him.

When a chance arose, the Boss took me aside and said, 'That's a nice lad you've brought with you. A surprise, I must admit, but I'd like you to know that if you've any plans for marrying, I shall be very happy to have him at Aroa.'

I could see the drift of his thoughts: another worker, a strong, energetic young man. 'I've no intention of marrying anybody,' I told him decisively.

After a meal of crayfish, the men retired to sleep in the car and van while the Doctor's wife and I made ourselves beds amid the petrol cans and fishing tackle.

Ben ferried us all to the island in the morning, enabling Jack a brief glimpse of my new home. The Doctor was occupied selecting the cattle he wanted to keep while I haggled with the Irishman for the house-cow, a few hens and the shearing machine. Before his departure, he explained how to work the boat-winch, the quirks of the water supply and various matters relating to the farm. Then I returned to the mainland to construct a temporary pen at the top of the beach to hold my sheep. They were to arrive early next day in two big lorries and Ben had agreed to help me transport them across the water.

After lunch the Boss, Jack and I climbed the hill behind the steading to get a view of Aroa. A path led up through the fir plantation and out along a steep-sided headland. We could look back to the high core of the island from this point and down into the wide sandy bay of Waiiti on our left and to the pebbled Home Bay on our right. It was a blue and white day, small clouds and enough breeze to fleck the sea. Shadows moved over the flanks of the hills, patterning the grass in yellow and green, darkening the gullies, thickening the water. We could see lazy waves breaking in a scalloped line on the wet, rippled sand and peer over the cliff edge to watch gliding ghosts of fish weaving between the tossing forests of kelp which grew on the seabed. A trawler steamed on the horizon and gulls banked and dived, their plumage caught in the sunlight.

The Doctor looked about him and gloated. 'Have you ever seen anything as beautiful?' Obviously he expected no reply, for presently he began to quote:

'Come my friends,
'Tis not too late to seek a newer world.
Push off and sitting well in order, smite
The sounding furrows; for my purpose holds
To sail beyond the sunset and the baths
Of all the western stars, until I die.
It may be that the gulfs will wash us down:
It may be we shall touch the Happy Isles.'

Jack and I glanced at each other, amused. Intercepting the look, the Doctor said sharply. 'Well? Where does that come from? Don't you know?' And when we shook our heads. 'Ulysses! Ulysses! Pommies and you don't know Tennyson.'

We were saved from making any guilty excuses by the sight of Ben's boat rounding the reef. It had been arranged that he would call for Jack on his way back to the mainland from the fishing grounds where he had spent the day. Reluctantly we went down to the harbour to meet him.

Valiantly Jack wished me good luck. 'No second thoughts?' he asked at the last minute.

But I shook my head. Nothing would have induced me to change my plans now.

The Irishman and his family left the island the day after we arrived. The crucial factor was the weather. If a wind rose and the sea became too rough for the small open boats, I would be confronted with the problem of how to feed and water the sheep. Once released from captivity it would be no easy matter to gather them all together again from the scrubby unfenced country around the Maori fishing settlement. I kept studying the sky, measuring the waves, testing the wind, and that night I woke several times and knelt up on the bed to peer anxiously out of the window, trying to make out the character of the sea in the darkness.

I was lucky. Dawn revealed a peaceful expanse of water

and a light, offshore breeze. Ben had mustered his friends and relatives to help us. The women and children had been sent to pick the long, tough leaves of flax bushes, which they ripped into strands, ready to bind the sheep's feet, a fore and hind together. Boys with horses and sledges waited to take three or four sheep down the beach from the pen to the boats, where the men lifted them aboard, packing fifteen or twenty in at a time.

I accompanied Ben with each boatload, to count the ewes as they were liberated and see them flocking up the beach to the green grass. Between journeys I worked in the pen, alongside the Boss and a couple of volunteers, catching sheep and tying their legs with flax. It was hot and tiring, and hard on the muscles of the back. There was no question of relaxing. I was in a fever to get the sheep across to Aroa before dark and before a wind should ruffle the sea. The loaded boats sank so low in the water that it would not have taken much of a wave to break over the gunwale. Towards midday, the breeze did freshen and I watched with dismay as the sea become corrugated with waves, but as the afternoon progressed, the air quietened again.

Towards evening, I realised that however frantically we worked we would not be able to get all the sheep to the island before dark. There is no long twilight in the north of New Zealand, and while we were tying the legs of the last thirty sheep, darkness fell. It was impossible to see to knot the flax and although I wanted to work by torchlight I was compelled to agree with Ben that it would be wiser to wait until the morning. It would have been risky to land the sheep in the dark and be sure that none had turned back into the water and got out of their depth, as several had done during the day. Indeed, we had nearly lost one ewe, who proved to be blind and had swum out to sea instead of ashore. So, reluctantly, I cut the flax bonds which we had just tied and left the last of the flock in the open for the night, with instructions to Ben to release them if rough weather should prevent me returning for them next day. Then, thanking my helpers, I followed the Doctor down the beach to his boat. The noise of the engine did not encourage us to talk.

As we drew nearer the island, he stopped humming and tossed his cigarette stub into the water. He sat up straighter, his head moving about as he peered in various directions, then leaning over he adjusted the tiller. He thought that I was steering too close to the reef.

His wife came down the beach to greet us, a lantern in one hand. She had an excellent meal prepared for us in the house. 'I hope it's not spoilt,' she said, 'it's been ready some time.'

We were very hungry and appreciated everything offered to us. Afterwards the Doctor brought out a whisky bottle. 'Just a nightcap to make us sleep.' The fact that I was drooping at the table, hardly able to keep awake, had bypassed his attention.

He began to talk at me, rather than with me, about farming matters. At length a sudden silence prodded me to life. He was sitting on the opposite side of the table, observing me with compassion. 'Poor child. You're exhausted,' he said. 'Don't stay listening to me. Go to bed.'

He lit a candle for me and I went to my room. A quarter of an hour later he tapped at my bedroom door and looking in, announced, 'I've come to tuck you up.' And proceeded to do so smiling benignly. 'I don't want you to feel homesick,' he continued. 'I want you to be happy here.'

I watched him in amazement. What a very strange farming partnership this was!

The calm weather held and I was able to bring the rest of my sheep over to the island the next morning. As soon as they reached the grass at the top of the smooth, wide sands of Waiiti where we unloaded them, their heads went down. They had not eaten for nearly two days and they were ravenous. I heard them tearing at the grass and they did not look back once, but nibbled their way inland along the green banks of the creek. I felt lightheaded with relief and joy to see my sheep safe in their new pastures. Never had I possessed anything so valuable and wonderful. I was as proud as an Australian grazier with a flock forty thousand strong.

At first the sheep seemed apprehensive of their new surroundings and made little attempt to scatter and explore, grazing in a tight mob. Thus they were easy to muster, and deciding to take advantage of this I drove them down to the yards a couple of days after their arrival to pare their hooves. It was tedious work: two hundred ewes, each with eight cleats. I had a pair of cutters like secateurs and only the Boss to help me. Rather unkindly I set a hot pace, ready to punish myself to keep my companion moving, knowing the last thing any man likes is to look weak beside a girl. When the back-aching job was finished, I had to admit that my landlord, old though he was in my eyes, could not be condemned for lack of guts.

Long after dark he was out with a Tilley lantern feeding his pheasants, planting seedling trees, or hammering in the workshop. Indeed, nearly every night when I tried to go to sleep, I would hear the incessant thud of a hammer, or the wheeze of a saw. On one occasion, past midnight, he knocked on my bedroom door to ask if he might fix a bolt on it. He was busy putting locks and bolts on everything for my security.

One might have thought that all this activity would make him sleep late in the mornings, but he rose hours before dawn. Round about five o'clock I would be disturbed by his feet crunching on the pebbles outside my window while he inspected his fishing lines. Then after a great clatter of crockery in the kitchen, he would appear with an offering of early-morning tea, one might almost call it night tea, which I accepted with bleary half-hearted gratitude, repressing a temptation to hurl the cup and its contents back at him.

I longed for his departure. Until he had gone I could not even feel that I was on an island. There was no time or peace to explore. I needed leisure and quiet to look and listen and begin to know and love Aroa, but he wanted to see me 'comfortably settled' and day by day delayed his departure. After nearly a week his wife decided to return to the mainland to visit some friends who lived at a township about seventy miles away.

'So you'll have to put up with me,' my Boss announced breezily. 'But we'll manage: I'm a good hand with a frying pan.'

On the first day we felled a tree. One of the bridges over a stream threatened to collapse and in order to repair this, we needed strong timber. The Doctor had already selected a suitable tree near the heart of the island and while he set forth on foot I rode one of the horses with tools, picnic and billycan slung in sacks across the saddle.

There was a magnificent view from the hillside where we worked. Half the island lay before us, golden and green, undulating down to the vivid blue sea, the sweeping sands of Waiiti and the dark feelers of rock. Trees grew behind us and fantails fluttered and called amongst the foliage. It was warm, and we could see my sheep resting near the shore under the vast sunshade of a pohutukawa. The Boss was like a boy on holiday, feeding sticks to his fire, boiling the billy, throwing bits of sandwich to the birds and lying back in the sun for a smoke.

We worked at the tree intermittently for hour upon hour. I never imagined it would take so long to fell. Backwards and forwards, backwards and forwards we drew the

94

double-ended saw until our hands were blistered. And when eventually the tree crashed down, there were still all the branches to be lopped. Even after the sun had set, we continued working. A bright half-moon cast a silvery path across the water and a few yellow stars from the Maori settlement winked along the mainland shore. The Doctor quoted from Tennyson again.

'The lights begin to twinkle from the rocks,
The long day wanes, the slow moon climbs,
The deep moans round with many voices.'

Then he found a white, star-like flower on some kind of native creeper festooning a tree and insisted I wore it in my hair like a true Polynesian. 'I'm a sentimentalist,' he said.

After supper back at the homestead, we walked along the shore. The Doctor wanted to find a crayfish. It was ebb tide and the night very quiet and still with countless stars in the sky, the moon haloed and a gentle surge lapping over the pebbles. The water felt cool to my bare feet and everything glimmered mysteriously in the half-light. The rocks took on strange shapes, crabs fled noisily and fish splashed wildly in land-locked pools. I went on ahead of my companion who lingered to prise open oysters with his penknife and poke into crevices with a stick, hoping to find a lobster.

The low tide made it possible to climb over rocks usually submerged and reach the outermost point of the headland dividing the Home Bay from Waiiti. There I sat down on a boulder and looked around with rapture. The long, undulating line of the mainland hills appeared black against the blue-black sky and the sprinkling of yellow lights had been extinguished. Isolated rocks off the point stood like nine-pins grouped at random, while the water slapped and slithered into the clefts and crannies of the reef. Behind me the island lay motionless, humped and curled up like an animal asleep.

But during the night the weather changed unexpectedly and by morning it was raining and blowing. I overslept and came into the kitchen to find the fire lit, the cow milked, the hens fed and the Doctor busy frying bacon. I sat down to eat guiltily and when it became apparent that we were running short of bread, I offered to try my hand at baking a couple of loaves. Delighted with my suggestion, the Boss went out to chop wood, returning at intervals to stoke the ravenous stove. I had never made bread before and the result of my labour was disappointing. The loaves emerged from the oven very flat and hard. They were more like vintage scones than bread, with a curiously potent yeasty flavour, an original taste indeed. I was annoyed: such a waste of time and effort, but the Doctor seemed glad of an excuse to tease me and at every meal he hacked off slice after slice, munching heroically. 'It grows on you,' he said.

That afternoon I rode out round my sheep. I liked to look at them daily for the flock was not settling down without casualties. Already a ewe had drowned in a ditch and a ram had got trapped on a cliff ledge from which he later fell, but miraculously escaped any visible injury, though I doubted whether he would be much use as a sire. Also, I wanted to gain the upper hand of my dogs. They were so brimming with energy and eagerness to work that I found them difficult to control, but the importance of doing so was brought home to me by more than one near-catastrophe when they drove mobs of sheep towards the cliffs. I was hampered by not knowing the words and whistles to which Glen was accustomed, while the puppy, Jock, though willing and keen, had never been trained. They needed discipline and I could see that it would take weeks of patient work before they began to be more use than trouble.

The next day did not pass quite so smoothly. As it was

dry, we worked to repair the bridge, but the Boss came to the conclusion that, like most women, I was no use with tools. He was quite right about my lack of skill, as I demonstrated to my consternation that afternoon. In swinging an axe, (perhaps he should have been using it himself instead of standing by) it somehow deviated on its course and caught him on the thigh. Fortunately it produced no mortal wound, but he hopped and gasped and a small patch of blood appeared on his trousers.

'Gosh! I'm terribly sorry,' I said, very worried for a moment.

He retired behind a flax bush to examine his injury.

'It's only a graze,' he called out valiantly, but he sent me down to the house for a bandage and penicillin.

When I returned he was sitting on a log with his trousers rolled up. The cut was about an inch wide and not very deep. My concern rapidly diminished as he instructed me, with a grin of enjoyment, how to bandage his leg.

That evening his wife returned and I learnt of their imminent departure. The Doctor's medical work could be postponed no longer. He rushed round making everything safe and easier for me: more bolts on doors and windows, the various engines oiled, greased and ready to start, a list of 'little tasks' drawn up for me and finally he handed me a rifle and an automatic pistol. A cow might fall down a cliff, he explained, and have to be put out of its misery. Besides, one never knew when one might need protection. I should not be able to call upon neighbours or the police if a sudden crisis arose. I hid the guns in my bedroom. If only we could have foreseen future events, I should never have been given such weapons.

The dangers of living on the island had been explained to me by everyone. The Irishman and his sons had taken a delight in telling me of shipwrecks, ghosts, sheep rustling and all manner of devilish sea creatures from sharks to stingrays and deadly jellyfish. Local stock agents had warned me of the hazards of farming such a rugged and inaccessible property, and now the Boss was turning the house into a fortress, cluttering the kitchen shelves with boxes of ammunition and medicaments. He had also arranged an emergency signal with the Maoris on the mainland: at night, a lamp was to be shone repeatedly against the white doors of the island boatshed or, by day, a fire lit on the beach with plenty of billowing smoke. It was to be Ben's responsibility either to come himself, or to send someone else in answer to my SOS. He had also agreed to bring over to Aroa my stores and mail once a fortnight, weather permitting. My landlord's twelve-foot dinghy had been built for inland waters and I had promised not to use it, except on very calm days, but only the roughest of seas would keep Ben away.

On the morning of departure, as the sea was quiet and perhaps with my cooking and bread in mind, The Doctor persuaded me to take a trip to the mainland. 'Have lunch with us,' he suggested. 'It may be your last square meal for months! I know you can't get rid of me fast enough, but I'd like you to join us in a little celebration . . . a toast to your future.'

So we drove about thirty miles to find a hotel where we ate an excellent meal and drank liberally. The Doctor never stopped talking, pouring upon me a spate of instructions: must wear a life jacket in the boat, never climb on the cliffs alone, oil tools after use because of the rusting effect of salt air, keep the horses exercised, write regularly . . . I smiled at his fussy anxiety, supremely self-confident in my total ignorance. By the time we got round to shaking hands and saying goodbye, I was afire with impatience to be gone.

I returned in the island dinghy to Aroa. Alone at last! Immediately the world was a different place. Now there

98

would be time and peace. Now I should find out what the island and an island life were really like. Steering across the blue open sea, I felt intoxicated with something much stronger than the wine we had drunk at lunch.

The dogs were waiting on the beach and flung themselves at me when I jumped ashore. It was a good homecoming; the water lay calm in the bay, the winch started easily and I drew the dinghy up the steep shelf of pebbles into the boat-shed successfully. Then I went into the house to change my clothes before milking the cow. As I entered my room, I noticed a posy of wild flowers arranged in a shell on the dressing table. A parting token from my landlord, who must have put it there when I was launching the boat. He astonished me.

When I awoke in the mornings, I lay listening to the waves on the shore and watched the paling sky through the small square window. If surf pounded or rain hammered on the low tin roof of the house, I shut my eyes again, but if the sky was bright, I began to stir and leaning out of bed, opened a door which led directly into the garden. Through this came the first shafts of sunlight, and the smell of soil, wet grass, seaweed and rocks. The dogs would creep in guiltily and curl up together on the sheepskin rug. Possessing no clock, I learnt to judge the time by the pattern of sunshine on my bedroom wall, but time was of little importance. Eventually, I would kneel up and look out of the window at the sea. It was almost like being in a ship's cabin, so close to the water that spray lashed the glass in a southerly gale and spring tides built banks of pebbles against the outer wall of the house. If the sea lay calm and tantalising in the bay, I flung off my nightdress and went down the beach to swim.

I never swam out very far, in case of sharks, but it was safe enough in the shallows. The water was so buoyant that I could float without moving my arms and legs and when no wind brushed the surface, every shell and stone and fish and ripple in the fields of sand was visible. Glen always swam with me as he loved the water, but Jock sat on the beach, a worried spectator, only moving to chase away audacious seagulls. The island and its spoke-like reefs abounded with every kind of seabird: shag and shearwater, tern and petrel, gannet and gull. Often, in the early mornings, flocks of gulls and shags congregated noisily on the rocks before flying out to fish, but the gannets, most powerful and beautiful of all, seemed to like fishing alone and would suddenly drop from the sky like white stones upon their underwater victims.

After swimming I dressed. Shorts and a shirt, or a cotton dress, never any shoes, never any coat, and a pullover only in the evenings or at sea. Such light clothing gave a feeling of freedom and agility. It was easy to run, or jump into a boat, or ride the horses bareback, or upturn a sheep. I had never cared much what I wore and now I cared even less. I thrived on the sun and sea air, becoming very brown and fit, sleeping well, eating well, feeling well. Illness was something I never considered. It did not exist.

Before breakfast I fed the ducks and hens, and milked Blacky, my £9 cow. She was an elderly, maternal creature, very docile and affectionate. When called, she would come to meet me across the ten-acre paddock by the homestead, where she lived all the year round, and would stand to be milked at the point where we converged, turning her head occasionally to lick my arm with her warm, rough tongue. Then I would put the bucket of milk to cool in the cold water of the stream.

I was not domesticated and never wasted much time indoors, so the simplicity of the house suited me. The walls inside were unpainted wood and there were numerous shelves for books and a miscellany of objects: fishing hooks, lures, lamp mantles and sheep raddle. Stockwhips, bridles and oilskins hung in the porch. Bunches of drying maize dangled like orange chandeliers from the kitchen ceiling, where a circular patch was darkened by smoke from the lantern which was hooked there night after night. There was a big deal table with benches round it and an ancient and inefficient cooking stove, which voraciously consumed firewood. No sink, no sanitation, no electricity, no radio. Outside was a cold-water tap, which supplied a flow of rusty water and an occasional eel, a wash-house with a copper, and a spring in the paddock for drinking water.

Most of the furniture had been left by the outgoing family and in the sitting room the chairs were shabby, but deep and comfortable. There was a good wide hearth, with a chain for hanging a heavy kettle, and though fires were only necessary in winter, the early dusk and plentiful driftwood often tempted me to enjoy a blaze. An ancient upright piano stood in one corner and a legacy of blood-and-thunder paperbacks collected dust on a high shelf. A chart of the island and surrounding waters hung upon one wall; and there were scallop shells on every ledge and sill, ideal receptacles for candles and sweet papers, buttons, nails and fish-hooks.

The house had three bedrooms, two of which had been added on the inland side and could only be reached through the garden. They had wooden bunks built into the walls and each could accommodate three people. The third room, which was mine, looked out over the sea and was half-filled by a single bed. It was the smallest, quietest and pleasantest corner of the house. A room on the shore.

My menu was wholesome, but unambitious. Poached eggs and a glass of milk. Fried fish and kumaras. Oysters off the shell and watercress from the stream. Accidental mutton, eaten in mourning. And figs or peaches, melon or passion fruit for dessert, with cream. Always cream, fresh bowls,

101

and sour bowls maturing into cheese. And then there was my bread, brown, very brown or black. I could never be sure until I took it from the oven whether it would be praise-worthy, just edible, or fit only for the hens.

During my second week on the island, I found a vine of big, purple, overripe grapes and started making wine, with-out knowing how. After a few days, the liquid began to bubble and I was filled with great expectations, but during the following months, it lost its kick and tasted disappoint-ingly vinegary.

I had a lot to learn, and the Maoris provided useful clues to economical sea-shore living. They knew when the sea was safe, the best places to fish, where to find bait and to set traps for crayfish. They showed me how sea urchins could be cracked open and eaten raw, and how pauas or abalones could be extracted from their irides-cent blue-green shells and pounded, to tenderise them before cooking their mushroom-like flesh. They practised a simple method of preserving a big catch by hanging the gutted fish above a smouldering fire in a special smoke-house the size of a sentry-box. And they could tell me which plants were edible, including a wild root, taro, which proved an excellent substitute for potato. If a plate or hat or mats were required, the women could weave them on the spot from flax leaves. Sometimes, two or three of the fishing boats anchored in the bay, while the Maoris waded in the shallows netting piper or searched for shellfish on the reef, but they never ventured beyond the shore, except to come to the house with gifts of fish and kumaras. Often, on a calm morning, I heard their singing and laughter in the far distance as they sailed out to the hapuku grounds, which lay three miles beyond Aroa. But sometimes I saw no one for weeks, apart from Ben twice a month with stores and mail, and then I suspected the Maoris were celebrating. They lived to enjoy the present, working and spending alternately, and cheerfully growing fat with age.

Many years ago Maoris had inhabited the island. The terraced remains of ancient fortifications were still visible on a headland and the Irishman had shown me skulls, arrowheads and old tools which he had found. It was rumoured that less than 150 years back, there had been cannibalism and savage wars at the same settlement where Ben and his relatives now lived; very hard to believe when I considered the kindness of my neighbours. If they felt any resentment that Aroa was now in white hands they did not show it, but remained consistently friendly without ever intruding.

My days were seldom planned. The weather decided my activities, but usually in the mornings I rode round the stock, watchful for sheep caught on cliffs or stuck on their backs, for gates blown open or fences in need of repair. There were four large paddocks on the island and intermittently the cattle had to be driven from one to another, counted, and water supplies checked.

When the afternoons were devoted to work, a wide choice lay before me. I might take one of the draft horses and a sledge along the shore to gather driftwood, or clean the boat, or cut some of the manuka scrub, which encroached upon the pastures, or work in the garden. And then, of course there were the seasonal jobs with the stock: lambing, shearing, dipping, drafting cattle and ear-marking calves.

But sometimes I did not work. If the sun scorched my back and the sea tempted, I would go down to the shore and swim and lie on the sand, or launch the dinghy and explore the reefs and little islands. Glen loved a voyage and would sit like a figurehead in the bows, but Jock, if forced to accompany us, curled close to my feet, and was only happy when we reached home again. Sometimes it would rain peltingly hard, turning the streams brown with mud and the sea

103

round the island thick and milky, and then I would stay in the house to read or write letters or bake bread.

In the evenings I cooked my main meal. It was the only time when I cooked anything and I would let the dogs into the house for company. We were a quiet trio. I loved to sit listening to the soft breathing of the waves on a calm night and watch the flames of the fire licking round strange, bleached arms of wood. Or when it was rough and surf pounded on the beach and the shingle was sucked back with a screech, there was an even greater joy in the contrasting peace within the house, and the knowledge that the wild sea would permit no one to come. I loved the isolation.

Often when the night was warm and moonlit, I went for a walk before going to bed, through the firwood, up over the headland and down to the sands of Waiiti, the most beautiful place on Aroa. Indeed, at low tide, when the wide virginal beach gleamed wet and silvery, the surf shone ghostly white and the sharp, serrated teeth of the rocks loomed black, it was incomparable there. Always the rhythmical roll of the waves and a shrill chorus of cicadas from the grass bordering the shore. And perhaps the bleat of a sheep on the hill. And the strong salt smell of the sea.

That first spell of solitude was broken too soon for me to discover how long my happiness might have lasted. I could not believe that I should ever fail to recognise and love the beauty of Aroa, but I had never deluded myself that I should want to live alone forever. It was quite impossible to imagine myself at the island as an eccentric, white-haired crone, or even as a shrewd, hard-bitten, middle-aged farmer. I knew that it was essentially an adventure of youth: active, strenuous challenging, risky and romantic. How wide was the gulf between living alone by choice when young and coveted, and living alone through necessity when old and forgotten. But solitude had never been a goal in itself; it was simply a part of the pastoral environment which I loved. And already I had discovered the vital importance of letters, the pleasure of greeting Ben, the tendency to talk to animals and dote upon them when there were no people, and the need for being needed, if only by sheep. Caring for my flock was the sane, rewarding basis to my life, the point of it. My happiness might soon have faded if I had had nothing to do but sit and admire Aroa. Instead, my work made me feel required and a part of the place.

And so I enjoyed living alone, the tranquillity, and work, and the idleness. But one Sunday, looking down from the headland into Waiiti Bay, I saw a yacht anchored there and my excitement was like that of a child, intense and uncontrolled. For a moment, I was held to the spot by the loveliness of the white-hulled ship swinging in the surge. Her slender mast and folded sails and trim, clean lines gave her the beauty of a seabird resting on the water, and I could imagine her exquisite grace and speed in action, with the wind swelling her sails. I started to run down the hillside, the dogs bounding ahead infected by my eagerness. Then they began to bark madly and I heard voices placating them on the beach.

I slowed down and hesitated, feeling suddenly shy as I came within sight of a group of people sitting on the sand. I was disappointed, almost antagonistic. Picnickers! Tourists! No doubt I had been unconsciously hoping to find a solitary, bronzed Adonis. But it was too late to turn tail. Jock had already placed himself in front of a woman.

The woman smiled. 'We heard there was a girl farming Aroa.' They all looked at me. I felt like a creature in a zoo.

'Are you really on your own? You must be brave!' they said.

I sidled over to Glen and booted him away from their bag.

105

They invited me to share their meal and when I saw that they were eating cold chicken I accepted with alacrity.

It took me several minutes to figure out who was the skipper. Besides the woman and two young boys of about twelve and fourteen, there were three men, one old, one middle-aged and one young. They were out for a day's sailing and had come about ten miles up the coast from a small settlement on a river estuary. But when I began to ask more specific questions about the yacht and turned to the oldest man for an answer, he said, 'You must ask Bob,' and indicated the youngest of the trio.

He was lean, sallow and black-haired, maybe three or four years older than myself. His manner was unassuming and he spoke slowly and quietly, while his eyes flickered briefly upon the person he addressed before looking down at the sand, or away at the sea. I learnt that he was a boat-builder and had constructed the yacht in his spare time over a period of three and a half years. I might have guessed then that he was a patient sort of person. Nothing ruffled him. Nothing made him hurry.

When we had finished eating, the skipper offered to show me over his boat and dragging a small dinghy into the water, prepared to row. At this point, Glen took a flying leap amidships, nearly knocking us into the sea, and began to shake his wet coat vigorously. I shoved him overboard hastily, aware of my companion wincing as the dog's claws scrabbled on the virginal paint. Speedily we pulled away from the shore.

The yacht was superbly built and fitted, without being large or luxurious. Twenty-two feet in length and containing one roomy cabin with two bunks and a small bunk in the foc'sle. Everything looked very bright and beautifully cared for: even the stainless-steel sink and the stove on its swivels shone, and one could see one's face in the central table. It was almost too tidy, all possessions stowed away in

106

the lockers under the bunks and the crockery anchored in special niches within a cupboard so that nothing could slip and get broken in rough weather. On deck the orderliness was equally apparent; ropes, cleats, planking, hatches, appeared so clean that one hardly dared to tread, even with bare feet washed by the sea. True it was a new boat, but it sparkled with the meticulous finish which only a skilled craftsman would slowly and lovingly add.

I was filled with admiration and envy. It seemed to me the most wonderful achievement to produce from strips of wood this work of art. Here was the very antithesis of the botched factory job. This was the realisation of a whole-hearted ambition through knowledge, perseverance and unnoticed, unmeasured overtime.

'Can you sail her alone?' I asked.

'Oh yes, she's quite easy to handle.'

'I'd like to learn how to sail and have a boat of my own one day.' It was a new dream, born that afternoon.

'Would you?' he said, looking at me speculatively.

We talked of other things and presently went up on deck to return to the shore. But while he was untying the painter of the dinghy, he said suddenly. 'I could teach you what I know about sailing, if you like.'

When we rejoined the others, they all accompanied me over the hill to the house, where I made them a cup of tea and exchanged figs for newspapers. Having no wireless, I tended to lose touch with the world, but the news was much the same as a month, or a year ago: crises, riots, discord, all so far away, unreal and baffling. Peace seemed so easy to find, if one searched in the right place. Not only my island, but the whole of New Zealand and a thousand small, remote countries could still doze while the lions roared.

Bob returned the following Sunday and we went sailing. The sea had been rough during the week and there was still a swell running, but the sun shone and a fresh wind blew.

107

We circled the island and it enabled me to see the entire coast for the first time. I was amazed to discover how rugged Aroa appeared. Sailing in the dark shadow of precipitous cliffs, skirting jagged reefs where the swell broke in clouds of spray, tacking between tiny, rocky islets, the wildness of the place took one's breath. I could hardly believe that this mound of land, girded by boulders and sea spume, was my home. Apart from the gulls and terns and the occasional white speck of a sheep, the island might have been uninhabited. The beaches lay deserted, sands stretching naked, without a single footprint, and in the cliff faces, black caves gaped like empty mouths. I believe I was a little frightened. Certainly, I was nervous of sailing so near the rocks and sometimes seeing within a stone's throw the water turn pale over a submerged reef. The Maoris had warned me of the dangers of taking a boat round Aroa, but although they had described the worst hazards, I did not know my bearings sufficiently well yet to be able to pinpoint a hidden rock under the treacherous expanse of water. But Bob was calm and watchful and we met with no disasters.

We towed a spinner in our wake for kingfish and kahawai. And near the northern shore of the island we ran into incredible shoals of blue and red maumau. Translucent fins and tails whipped and churned the surface of the water until it frothed and bubbled. There must have been hundreds of thousands of fish, milling together, feeding on plankton, the two colours never mixing, but weaving coral and turquoise patterns upon the sea.

Later, we were chaperoned by nine porpoises. They leapt from the water, sleek and dark as seals, a pleasure to watch, so smooth and graceful and jubilant, reeling effortlessly through the air.

We found a sheltered bay in which to anchor for lunch, and caught about a dozen snapper while we sat on deck eating. They were biting well although the water was so clear that we could look down and see the fish approach our bait. I let Bob do most of the fishing and take my captives off the hooks, for I never really enjoyed using a line. And I asked him to knock each victim senseless, hating to witness fish jumping and flapping and gasping for air.

The wind had decreased to a light breeze during the day and in the afternoon I was allowed to take the helm. I was profoundly happy watching the sails, feeling the yacht gliding forward, listening to the creak of ropes and canvas and the swish of water as the bows thrust ahead. Bob stood by, watching for rocks, ready to take over from me if necessary. He was an ideal instructor: quiet and patient, never fussing, never ruffled.

We passed tantalising little islands, green, hummocky jungles of flax and stunted trees, inhabited only by birds and a few wild goats. And there were strange outcrops of rock; one like a hand with a finger pointing upwards, another forming a perfect arch, through which a small boat could sail, and tall stacks and pinnacles which might have broken off from Aroa long, long ago. We saw a shag colony and a whale spouting far out at sea. And in the late afternoon, we sailed past the lagoon-like bay of Waiiti, rounded the headland and came back within sight of the house on the shore. Although the yacht moved so quietly through the water, the dogs must have heard us for they raced down the steep shelf of pebbles and barked with joy.

Like myself, Bob was a stranger in the district, having come to live there temporarily while building a boat for a fishing lodge, and so the coastal waters were new and exciting to us both. And this Sunday proved to be the first of many such expeditions through which we gradually became familiar with every island and beach and reef for miles. When it was warm, we would find a sandy bay and swim, and Bob, who possessed an aqualung, would dive to admire the underworld and search for a crayfish for supper, while I watched his path of bubbles anxiously. Then, in the evening, we returned to Aroa and after milking the cow I would cook some of the fish we had caught, while Bob repaired whatever engine or object I had ruined since his last visit.

Often he stayed the night at the island, anchoring in the bay, and I could look out of my window and see the ship's lights bobbing on the water. But if the weather looked ominous he set sail, guided in the dark by the long dim silhouette of the mainland hills, always blacker than the blue-black sky.

We must have spent half a dozen days together when he began to talk of building me a boat. Something small, light and easy to handle with a sail. My cup was full.

A few oaks planted round the homestead had turned golden, but their leaves had not yet fallen. The native trees never lost their foliage or their dark-green colouring. On sunny days it was still warm enough to swim and there was not a hint of winter in the garden. Strangely, English flowers like roses and lavender continued to bloom, but a row of pineapples reminded me of the true climate. In the moist gullies I found mushrooms six inches in diameter, and the figs were ripe for picking.

There were, of course, minor disasters to prick me when I became too cocksure. Three sheep had died, two over cliffs and one from unknown causes. The boat-winch went wrong and after cleaning the sparking-plugs on the engine, I was flummoxed. The Huntaway pup, Jock, tore himself on a strand of barbed wire so that I was forced to put a stitch in him with a needle and surgical thread from the Doctor's first-aid box. When I attempted to wash sheets in the copper, they emerged from the water stained orange with rust, and the first time I went out with a horse and sledge for firewood, the load capsized twice while I was driving down the steep incline. Then there was the anticlimax, rather than disaster, of those spells of rough weather which kept Ben away.

On the appointed day, I examined the sky and watched the sea from dawn onwards. I could hardly eat my breakfast but kept jumping up to look out of the window, imagining that I heard the hum of a distant outboard engine, or perhaps mistaking a far-off seagull bobbing on the waves for the white hull of Ben's boat. No work was done on those days. Before Ben arrived, I could concentrate on nothing and after he had gone there were so many letters and papers to read; then I would unpack the food and enjoy the best meal since the last stores came.

Often Ben was on his way out to sea to fish, but he would always wait long enough for me to answer any urgent messages, or telegrams. I liked Ben. He wore such a wide smile and everything he said, he said as if he meant it. He would come into the house, shouldering my sack of stores, like a fresh breeze, saying perhaps of the sea in his explosive way, 'By Joves, she's wet today!' Then he would ask very politely how I was and we would talk of fishing and the island and the Maoris.

When he had gone, I knew that it was improbable that I should speak to another human being for two or more weeks, but I was too immersed in my life to care.

The warm, wonderful autumn cooled into winter. Torrential rain showers soaked deep into the summer-dried earth and the grass grew thick and green. My ewes were now in lamb and grazed the pastures hungrily, becoming rounder and fatter each week. Too fat. They fell into ditches and rolled on their backs and the number of sheepskins curing on the rafters in the woolshed rose to eight. I hated the grisly job of cutting up carcasses but the meat was too valuable as dog food to waste, while the skins could be sold or used, when cured, as rugs under saddles or mats on the floor. It is much easier to skin a warm sheep than one long dead, for the skin is still pliant. I would hang the carcass up from a rafter in the woolshed, hauling it there on a rope, cut the skin around the top of the hind leg with a sharp knife, then tug and peel the whole skin back down the body to the neck, just as I did when skinning a lamb for fostering. Any lumps of fat which clung to the skin had to be scraped off laboriously. Finally the skin was stretched tight and exposed to the salt air until ready for further rubbing down and drying in the woolshed.

One morning Ben brought me a telegram from the Doctor. It was like a thunderbolt on a calm, fine day. I read it and reread it. I was stunned and then angry.

'Dam builder arriving Saturday,' ran the message. 'Do not know him well, but believe him to be a reputable character.'

It was Thursday. Probably the man was already on his way north. The Doctor had often talked of creating a lake for wild duck and trout at Aroa by constructing a dam across the main creek, but I had never expected his plans to be put into practice so soon. We had never even discussed the feasibility of such a project, yet now he was sending a complete stranger to live in my home without consulting me at all. I was furious. Didn't he realise that he was endangering the whole point of my coming to the island? Instead of living in a free and peaceful world I might find myself in a prison along with another inmate whom I could neither avoid, nor tolerate. Heaven could change to hell in a day if we hated one another.

Fortunately, I had all Friday to calm down. A fine day, blue sea, blue sky, a gallop across the sands, and good work from the dogs put me in a brighter mood. My optimism revived. Perhaps the Boss was sending someone young and eligible, hoping to marry me off and obtain two workers for the price of one. But even if the man proved to be middle-aged, married and ugly, at least he would be strong and therefore capable of castrating calves, digging ditches and performing all the heaviest jobs on the farm. The more I thought about it, the more I realised that I could put a man to good use. I felt almost sorry for him in advance. He wouldn't stand a chance.

On the Saturday morning my confidence evaporated. I was in a last-minute panic and when the hum of a distant outboard motor sent me flying to the window, I wished to God that the boat heading into the bay would sink before my eyes. I wanted nobody to come. The serenity and loveliness of my first autumn on Aroa could not possibly continue through the winter.

I went down the beach, hackles up, but one look at the man standing waving in the bows of the boat, dispersed both my worst fears and most extravagant daydreams. Here was somebody who ought never to have come. As the boat drove on to the beach with a crunch of pebbles, he hailed me with a familiar 'Hi!' and tossed a small grey kitten into my arms. 'Present from the Doc.' Then he jumped ashore.

I hardly knew whether to laugh, or cry, or gnash my teeth with rage at the sight of Jack. Apparently he had talked the Doctor into offering him work. Though he knew nothing of dam building, he had swept aside all difficulties and constructed the dam so vividly in his own imagination and that of his audience that there had been no room for doubt. His eloquent tongue and powers of persuasion had proved irresistible. Listening to his explanation with uneasy misgivings, I very much doubted whether Jack could build a dam if he tried. Was he going to start digging it manually with a spade? I was sceptical about the whole plan, and saw it as an excuse to invade the island. He had seen my dream of living and farming at Aroa come true and was envious. I could not blame him. The unique beauty of the place and

113

the chance to swim, sail, ride, fish and live as one pleased, must have made the interminable round of Taranaki dairy farms seem flat indeed. I could have forgiven him if he had asked to come, but to give me no warning, allow me no choice and simply arrive on the beach like a shipwrecked sailor, made me mutinous.

Sexual equality and Women's Lib were unheard of in those days, but I was prepared to fight for my freedom. Thank God for the Doctor's bolt on my bedroom door and the abysmal standard of my cooking! Before the first week was over, Jack had learnt to bake bread, gut and fry fish, milk the cow and chop wood for the stove. The dam was never built; Jack decided the chosen site was unsuitable, but he painted the kitchen, put in a sink, ploughed up a piece of land for maize and helped with the stock work and ditching. We were energetic in spasms, lapsing into idleness when the sun and sea lured us to the shore, or into the boat to explore the waterways among the reefs and islets. Sometimes Jack took off to the mainland to get himself cigarettes. I never knew what to expect when I went down the beach to help him haul up the dinghy on his return. Once the boat was full of pullets. Another time he brought enough paint to decorate the entire house and charged it to the Doctor's account. And there were always peace-offerings of books and chocolates.

He was a turbulent companion. Sometimes the first thing he said at breakfast was 'Marry me!' Other mornings he might say nothing at all. Once he moved right out of the house and began cooking his meals over a camp-fire but soon he was back again, playing Bach on the ancient piano, filling the lamps, stoking the stove, whistling and singing. His spirits either soared or slumped. There was no comfortable medium. The calm, relaxed days, which I loved so much, had gone. Jack with see-sawing moods made me feel like someone at a fairground sickened by the roundabouts

114

and helter-skelters. Happiness was like a bubble which burst as soon as Jack tried to capture it. He was too impatient. His discontent festered like a sore in the sun and he waged a losing battle trying to convince me that marriage would be the answer to all my problems. That I had no problems, that I was perfectly happy living alone, that I wanted nothing and no one, were facts which he could not accept. I was adopting an unnatural life, he said. But it was my choice and all our arguments ended in this same deadlock. He could not influence me.

Eventually Jack's winter leave expired and he returned to his round of dairy farms in Taranaki. My landlord's matchmaking plans and hopes of a dam for his lake had come to nothing.

A flat calm lasting three days followed Jack's departure. The sea was like green glass, the mainland hills peered at their own reflection, and the tide ran out silently. Not a ripple disturbed the mill-pond water in front of the house, and looking down from the headland, I could see every detail of the seabed: the cruising fish, the waving fronds of kelp, and the gleaming white scallop shells lying on the sand. The breathless quiet was uncanny, for a day without wind was rare. Even solitude seemed strange after weeks of yo-yoing company.

On the evening of the third day, a long, heavy bank of cloud advanced from the west, and when I lit the kitchen lamp a swarm of seaweed flies and moths invaded the room. Hurriedly I closed all doors and windows as thousands of insects gathered on the panes, seething and humming like bees outside a hive. The air had become oppressively warm and sultry, and with relief I felt the first fanning of a breeze when I went out to shut up the hens. I slept uneasily, disturbed at intervals by the sound of the rising wind and strengthening waves.

115

The sea was transformed by morning from a limpid green lake into a heaving, tossing, wallowing turmoil of white-maned breakers. When I went outside to milk the cow, the wind greeted me in ferocious gusts, tugging at my hair and clothes and drenching me with sudden splattering attacks of salt spray. It was impossible to stand upright. In the garden young daffodils, just coming into bloom, had been laid and broken and petals were scattered like confetti under the peach trees. A forgotten sheet, which had maddened me all night flapping on the clothesline, had wrenched itself free and now hung over a rose bush, its edges ripped to ribbons by thorns. All the garden birds, fantails and finches and tuis, had vanished, flown no doubt to the protection of the firwood. And the sheep and cattle stood tails to the wind in the lee of the flax bushes and amid the sheltering manuka. The sky was thick with grey, fast-moving clouds with no sign of a rift, and the rain spat in squally outbursts. The mainland was obscured from sight and fountains of spray rose over the reefs.

All day I worked in the woolshed, cleaning up the pens, sorting and packing dags, scraping sheepskins. And later I set the grinding plate rotating, and sharpened the shearing combs and cutters while sparks flew like fireworks of golden rain. Even the dogs preferred to stay inside, crunching up hoof parings and flakes of hide, and collapsing with long sighs into the comfort of wool-filled sacks. The kitten had followed me too, stalking, pouncing and patting tufts of fleece blown along the floor in the draught. Outside, the surf roared on the beach, pounding and sucking back the pebbles. The storm showed no sign of abating. Indeed, as the day progressed it worsened, tearing at the doors and roof of the woolshed, funnelling under the slatted floor and lashing at the fig trees bordering the yards.

After dark, I heard the first distant rumbling of thunder.

When I let the dogs into the house, they crept under the table and I was tempted to join them there. But the tide was still coming in, and the wildness of the sea drew me to the window to watch the onslaught of waves rearing and crashing upon the beach, lit up by vivid snakes of lightning. It was far worse than the previous night; the whole house creaked and rocked in the fierce gusts of the storm, and I found myself holding my breath during each pause between the collapse of a breaker and the long drawn-out screech as the shingle was sucked back down the shelving beach.

I went to bed without much hope of sleeping and pressed the blankets round my ears, aware that the house was built far too near the sea. I was too frightened to relax, and lay taut, listening to the surf, measuring the size and strength of the waves by the time which they took to gather, and retreating deeper into the bed at every clap of thunder. Each gust of wind buffeting the house seemed stronger than the last; the window panes chattered, and a ducking branch drummed on the roof. Spray was flung like heavy rain against the wooden wall, which was the only partition between my head and the beach.

Now and then a wave, mightier than the rest, crashed down and sent water surging up the beach to overtake the line of weed and debris which marked the height of the last attack. Kneeling up at the window, peering through the salt-smeared pane, I tried to judge the distance between the house and sea, first in yards, then in feet, thinking all the time that the tide must surely turn soon. It was cold sitting up and I got back under the blankets feeling scared.

Presently a new sound made me snatch up my dressing gown. The surf had reached the house and was beginning to bombard the seaward wall with pebbles. I lit a candle and went into the kitchen. There too, the shingle clattered against the wall as each wave broke, and shreds of seaweed festooned the window. In the beam of a torch the sea looked dark and tumultuous, the crests seeming to rear higher than the house. Opening the back door I found on the path, just beyond the step, knots of kelp and a dead starfish, apparently tossed right over the roof by the wind. Returning to my room, I collected together a few valued possessions and took them to a back room of the house. Then I sat down at the kitchen table with a book and tried to read, but it was hard to concentrate. The dogs rested uneasily by my feet, rising from time to time and nuzzling me for reassurance. It was a long and terrifying night.

Towards dawn I returned to bed, sure at last that the tide was retreating. I slept heavily, exhausted by fear, and in the morning woke to find that the wind had spent itself and dwindled to an innocuous breeze. Though still wild and turbulent, the sea was gradually subsiding, leaving the beaches pounded into new shapes, with banks of sand and pebbles rising like ramparts, and every bay strewn with the loot of the storm.

I went out after breakfast with a sack to look for injured birds and gathered up a few dead fish for dog food. Along the shore lay great mounds of twisted seaweed and a line of driftwood, tree roots, branches, broken fish boxes, old oars and planks. Shells and fish were scattered everywhere: sprawling starfish, blubbery masses of jellyfish, open-mouthed glassy-eyed cod, snapper, maumau, brown leather-jackets and garish-red parrot fish. And here and there were dead birds, mostly the Little Blue Penguins which lived in the sea. The water, so churned in the night, looked thick and milky. Spray from the swell pounding against the rocks and cliffs rose in white plumes. Patches of blue appeared between the clouds, and shafts of sunlight threw pools of colour on the sea. It was beautiful but sad. So many creatures dead, so many plants flattened; even the grass would suffer, burnt by the salt spray. And all the time the giant waves reared, curled and crashed.

118

No one could come to the island, neither today, nor tomor-row. If a wind rose before the deeply furrowed sea had quietened, it might be a week or longer until a boat could cross from the mainland. Ben would not be able to bring my stores on Friday and there would be no letters.

Gradually I reaccustomed myself to solitude, and became utterly absorbed again in my work and environment. Towards the end of July, I went out each day hoping to find the first lamb. This was the time of year I loved best, hard and heart-breaking sometimes, but infinitely rewarding. The thrill of discovering new arrivals never grew stale. Birth remained a miracle. And so the first addition to the island flock caused great rejoicing.

A strong ewe lamb was born in the night and I found it next morning in a sheltered gully, curled in the rushes asleep, like Moses. Making the dogs lie down and slinging the horse's reins over a low manuka bush, I approached the lamb quietly and caught it before it had struggled to its feet. The curly yellow-white wool was still damp, but the lamb felt warm and lively in my hands. I looked to see its sex and to find out if it had suckled. The mother circled round me anxiously, nosing closer and closer, bleating continuously. Letting the lamb free, I stood back to watch it rising and collapsing, trying to master its brand new legs, hunting first at the wrong end of the sheep for milk, then coming at last to the right place and wriggling its tail in ecstasy.

Work was not always so pleasurable. Sometimes, when I went out in the half-light at five or six o'clock in the morn-ing, I found that the night had brought trouble: difficult lambings, dead lambs, weak lambs, abandoned twins. One day I came across a ewe bleating repeatedly in distress and showing every sign of having lambed, but though I hunted for two hours, there was not a trace of her lamb and I presumed it must have fallen into the sea. The next day I went back and the sheep was still bleating, so I searched

119

again but without result. And then, passing by three days later, Glen brought to my attention a deep fissure in the ground and there the lamb lay trapped and dead from starvation. The ewe was still grazing in the vicinity, but she no longer called.

Occasionally a young ewe deserted her lamb and had to be driven down to the steading and penned in with her neglected offspring. Weak lambs had to be warmed and bottle-fed in the house until they had gained strength. Others had to be taught to suckle, or fostered on to ewes which had lost their lambs by disguising them in their predecessors' skins. And then there were the deliveries, made difficult for me, because I had no one to hold the sheep while I examined the position of the lamb and struggled to extract it alive, usually on the bleakest hillside or cliff edge, a mile or more from the steading.

It was not an easy time, and when I had lost a ewe or lamb, or worked continuously through the daylight hours without finishing all that demanded my attention, I despaired. The island was too big, the land too rough. I could not get round all the sheep often enough. I tried to look everywhere twice each day, it was impossible to do more. But even then, I could miss a sheep or lamb in the scrub, or concealed in a hollow, or hidden on a green cliff terrace. I blamed myself for every disaster. It always seemed that if I had not been so blind, if I had gone out earlier, or returned sooner, given a lamb one more feed, delivered a lamb more skilfully, observed a ewe more carefully, every life might have been saved. I could not accept death calmly.

One evening I was walking round the sheep when I saw a ewe with her young lamb grazing near the edge of a dangerously steep cliff. I was afraid that the lamb, which was skipping and gambolling, might fall, so I climbed down to a lower ledge and whistled and shouted, hoping to drive the pair to safety. My mistake was to forget to keep the dogs

120

close behind me and Glen, eager to help, ran on ahead without my noticing, and following a higher path appeared above the ewe and began to bark excitedly. The startled sheep sprang away and ran down a short grassy slope to the rim of the cliff, just stopping herself in time. But the lamb, rushing helter-skelter after her, plunged over the brink before my eyes. I felt sick with horror and for an instant turned my head away. When I looked back, the lamb was struggling in the shadowy blue-black water, while the panic-stricken mother leapt down on to a rocky ledge and, peering at her offspring, bleated to it piteously. A heavy swell was running and surf crashed on the rocks below. I slid and slithered over the boulders down to the water's edge. Already the lamb, still fighting for life, was being swept out to sea by the current. I started wading after it, but the seabed was rocky and the wildness of the water frightened me, and when I had swum a few yards, I knew that I should never reach the lamb and fought my way back to the shore. Only one ear now showed above the water, like the white paper sail of a toy boat. I tried to climb along the face of the cliff, hoping to get nearer the drowning lamb, but it was no use, I could only look down at it helplessly and see its limbs waving feebly now, like white arms of seaweed. There was nothing I could do. The boat was too far away, the sea too rough and there was no one to help me.

Early next morning I went back to see if the ewe had got off the ledge, but she was stuck there and still bleating. Then I saw the dead lamb. The sea had calmed and the tide had driven it into shallow water. It was floating just under the surface a couple of yards from the shore and I could see the ghostly outline of a large fish, perhaps a shark, feeding on the carcass.

My attempts to get the ewe off the cliff failed, so I returned to the Home Bay and signalled for Ben by lighting a fire on the beach and heaping on wet kelp to make it smoke. He did not come and the next day it was too rough. Again I set out for the cliff, but this time with a rope. The sheep had not moved. Tying one end of the rope to a boulder at the top of the cliff, I knotted the other end round my waist and, petrified, started to descend. But this was not to be a lone heroic rescue. I gave up before I reached the ewe, afraid for myself and afraid that in panic the sheep would fall to a similar death as her lamb. Instead, I climbed back up the cliff, tore up an armful of grass and threw it down to her.

The following day the sea was slightly calmer, so I stoked up my fire for Ben and waited. He came in the afternoon, his boat wallowing in the waves, his small black-eyed grandson clinging to the thwart. I led the way to the cliff and thankfully let Ben take command. In amazement I watched him sprint down the cliff-face, bare feet clinging to the raw rocks, no rope, no hesitancy. When he was a few yards from the sheep, he called for me to lower an end of the rope and catching this, tied a slip-knot and lassoed the animal. Then, climbing down to her, he secured her in a sling and finally rejoined me to haul her to safety. I felt ashamed of my timidity and, wondering how to repay Ben, could not repress a bitter smile at the thought that the best thing I could give him would be a fat sheep for the next Maori feast.

During the lambing season I chose to ride a bay gelding called Skipper, the quietest and most cooperative of the three hacks. He would carry a couple of live or dead ewes across his back, or a saddlebag full of lambs. If I had to dismount to attend to a sheep, there was no need to tether him; he would graze nearby or stand meditatively until I was ready. He came when called to be caught, and would make his own way home if told. The only time he ever bucked me off was when I put a new saddle on him after three months of riding bareback. And he was a good cattle horse, quick to turn and energetic.

The next task was to tail and castrate the lambs. I used an

elastrator, a New Zealand tool which expanded small strong rubber rings so that they could be placed over the tail or testes to cut off the blood supply. The younger the lamb, the better, so there would be no check on its growth and only the briefest moment of pain. The scrotum and tail end which might otherwise get dirty and attract maggots, would gradually wither and drop off, causing no bloodshed or open wound. The method was much easier and cleaner than using a knife as we had done in Scotland and, by gripping the lambs between my knees, I could manage the job alone.

In the brief breathing spaces between the stockwork, I had the vegetable garden to plant and a patch of land to plough and work to a tilth for potatoes, seedling trees and corn for my poultry and the Doctor's pheasants. The draft horses had not been used sufficiently often and the younger mare was hard to hold, while the old one dragged cunningly behind, letting her companion do all the work. The plough itself was ancient and very heavy and I could not get the knack of guiding the team while holding both the lines and the implement. Finally I evolved a most unorthodox method, in which, having turned on the headland, I got the plough stuck firmly in the ground, then went round to the horses' heads, and tugging on one and dragging at the other, led them up the furrow. The plough worked much better by itself as long as it hit no boulders or tree roots. Cultivating and harrowing called for less skill. One could go in any direction, round and round, up and down or crisscross. It was a pleasant sort of job, especially with the breaks at the headland to give me a rest, as much as the horses.

We were working on a hillside, where it was possible to look over the pointed tips of the firs to the sea and back at the interior of the island, where the gullies and hill ridges ran down from the summit; the dark green of the bush contrasting with the young spring green of the valley grass and the red scarred landslides. All around the coast, the sea, grey or blue, white-capped or calm, washed against variegated rocks and sunburnt sand. The colours changed from day to day, hour to hour, bright and small-shadowed at noon, warm to the eye at evening, and misted and muted in the early morning.

CHAPTER NINE

Towards the end of August, when the lambing was nearly over, I received a message from the Doctor telling me that he intended to spend a few days on Aroa planting trees. Ben brought him across from the mainland one Friday afternoon. The fishing boat was weighed down with crates and sacks and baggage. We carried the gear up the beach. The sacks were full of seedling trees, and the crates full of beer and whisky which Ben eyed hopefully.

I wanted to go round my sheep before dark and the Doctor was eager to see how Aroa was looking. The Boss was in high spirits, pleased with everything he saw, admiring the pasture and raving over the view from each new vantage point. I believe he thought that I was blind to the scenery because I seldom spoke of it, yet the loveliness of the island was the whole reason for my being there.

When we came to the creek which ran down to Waiiti bay, seeing the watercress growing thickly there, the Doctor suggested that I cut a bunch for our supper as my feet were bare and I had a knife in my belt. I waded slowly downstream gathering the leafiest sprigs and then I noticed, a few yards ahead, a sheep lying half in and half out of the water. She was dead; so recently drowned that the flesh was still warm under her fleece. We dragged her on to the bank and I saw that she was heavily in lamb. She must have gone down to the creek to drink and the hollowed side of the bank had caved in beneath her weight. The Doctor seemed almost as distressed as I was, and volunteered to stay and skin the carcass while I continued round the flock.

A ewe was lambing on a small promontory jutting into the sea, and I was delayed some time waiting for the birth and then urging the sheep inland so that there would be no danger of the lamb falling into the sea. When I came back to the creek, the Boss was sitting smoking. He had given the ewe a post-mortem, hoping her lamb might still be alive, but instead he found twins not fully developed. I thanked him for his efforts and, knotting a couple of flax leaves, bound up the skin. I would fetch the carcass with a horse. Meanwhile we returned to the homestead feeling subdued.

Next morning the rain pelted furiously upon the tin roof, swirling down the guttering and beating in a million circles on the sea. I got drenched going round the sheep. A ewe had had twins in the night, but one had not suckled and felt cold and weak. I had to bribe her down to the ten-acre paddock, carrying the lambs in a bag and setting them in front of their mother at intervals to coax her forward. The sheep-paths were runnels of water, the grass squelched underfoot, and the rain ran down my neck. All round the island the sea was stained a murky brown from the precious topsoil which had been washed into the creeks.

When I reached the house with the lamb, I found the Doctor busy in his shirtsleeves, stoking the fire in the stove.

126

There was a faint smell of meat cooking and a pall of smoke from damp driftwood. He was roasting a leg of mutton from the sheep which had drowned. 'Too good to waste,' he said, but my appetite dwindled. However, I was glad to see him using the stove, for my experience of it had been frustrating. It seemed quite impossible to produce sufficient heat in the oven to cook anything in the usual way.

Having rubbed the lamb vigorously, I laid it near the stove, wrapped in a sack. It was lifeless, and the Boss watched anxiously while I tried to give it a little warm milk, first in a bottle and when that failed, from a teaspoon. The lamb appeared to have neither the strength nor the will to swallow, and the liquid spilled out of its mouth, while it threw back its neck and paddled the air with a forefoot.

'It'll die if it won't take milk,' I said.

The Doctor searched in his medical bag and found a stomach tube. 'Let me try with this,' he offered.

Very carefully, he administered a small quantity of milk heated to blood temperature. A few minutes later, the lamb raised its head and bleated. We looked at each other and smiled with optimism. Then I tucked the lamb up warmly and left it to sleep, while I changed my wet clothes. A quarter of an hour passed before I looked at the lamb again.

Its eyes stared up at me, unblinking. It made no movement. Hastily I unwrapped it from the sack and felt for the heartbeat. It was dead.

In spite of my protests, the Doctor insisted that it was his fault. He feared that he had fed the lamb too fast, or had given it too much milk. 'But I'll replace it,' he assured me. 'I'll give you a sheep for your birthday. Any breed you like.' He was so anxious to make amends.

It was not a successful day for him. Two hours after the joint had been put in the oven, the fat was only just melting. We ate bacon and eggs for lunch and sampled the roast, somewhat reluctantly on my part, at ten o'clock that night.

127

It was so tender that it fell to pieces on the plate, though the flavour was a little smoky. The Boss had had enough of cooking. He was resolved to purchase a new stove the moment he returned to the mainland. All day he had chopped driftwood to feed into the hungry fire, and the kitchen was like a furnace in spite of the open window and door; but the oven had remained obstinately lukewarm.

The rain never let up, and in the evening a thick sea fog shrouded the island. No wind blew and the muffled, leaden water in the bay made hardly a sound. Sitting in the living room, all we could hear was the intermittent dripping of rain from the overhanging peach tree, and the soft swish of the surge at high tide when it rose close to the house. While I wrote a letter, the Boss, who had collected a billycan of winkles before dusk, hooked them out of their shells with a pin and fed them alternately to the dogs, the kitten and himself. When they were finished, he roamed about the room, looking at my books on a shelf, at a chart of Aroa pinned on the wall, and his fishing line, which ran from the kitchen window into the sea. A tin can was attached to the sill and within this a pebble rattled if a fish struggled on the hook. The Boss was full of ingenious ideas, but he could never sit in peace, or in silence. I recognised his kindness, yet he reminded me of a buzzing bluebottle as he paced about the room, niggled by my self-sufficiency. Priming himself with whisky, he offered me a dram which I refused, then patting the sofa, assured me that I would be more comfortable sitting there. Embedded in a chair with a book, I shook my head emphatically.

Advancing to sit on the arm, he asked what I was reading, but hardly waiting for my reply, he launched into an outburst of praise. He praised my hair, my figure, my courage . . . my courage most of all.

'You lack nothing in attraction,' he continued, 'and I lack nothing in desire.'

Shocked and embarrassed, I shut my book. Then his hand reached out to touch my hair and I sprang to my feet. Escaping to my room, I secured the bolt which he had put there for my safety.

Next morning he signalled for Ben.

'You'll be celebrating when I've gone, won't you?' he said accusingly as he left for the mainland. I managed a wry grin. He was right.

I hoped to be left in peace after his recent behaviour but the Doctor returned too quickly. In November a telegram arrived to shake me from my summer siesta. He and Jack were on their way. They were coming for a Friday-to-Monday weekend with more seedling trees and pheasants. The announcement came as a shock. I had arranged to begin the shearing that Monday with the aid of a couple of men, and needed to muster the sheep on the Saturday and Sunday. Besides, twenty-four hours warning was not nearly long enough to catch up on the jobs I had postponed while celebrating the sun. It was difficult to know where to begin. I started working frantically mending and tidying the things most likely to hit the eye, beginning with the boat, which was full of sand and fish scales and remnants of rotten bait.

The day the visitors arrived, the sea was rough. It was a case of shouted greetings, then all hands to unload the cargo of young trees and crates of bedraggled birds while keeping the boat from being swamped by surf. The Boss lost a canvas shoe in the water and hopped on the shingle, yelling at us to retrieve it. Nobody took any notice. Ben was too intent on getting his boat emptied and heading for home before the breakers refilled it. Jack was flinging everything up the beach like one possessed while I hung onto the boat, endeavouring to stop it from swinging broadside onto the shore.

Eventually we pushed Ben's boat out through the surf and the Doctor reached for my hand. 'You'll have to help me. My feet aren't used to these stones like yours.'

Catching my reluctant fingers, he trod cautiously. Jack watched and bristled. It was not an auspicious start to the weekend, and indeed everything seemed to go from bad to worse. Apparently Jack had given up his herd-recording job in Taranaki and was being sued for breach of contract by the immigration authorities. The Doctor had taken pity on him and been harbouring him for several days. But Jack never brought peace in his pocket. He had not been on the island an hour before he was cornering me again on the threadbare subject of marriage. My exasperation grew in ratio to Jack's despair and the Doctor's agitation to get his trees planted. The men became antagonistic and abusive. Obviously, three was the wrong number for harmony.

Early on the Saturday morning we climbed a steep, pine-needled path in single file, heavily burdened with mattocks, spades and bundles of seedlings. I followed last and loitered deliberately to let the men start work ahead of me. I was fed up with them both, and watched with satisfaction first the Boss and then Jack bend their backs and move slowly up the staked rows, over the rise, and out of sight and earshot.

The freshness of a new morning was still in evidence: beads of dew lying in the leaves, cobwebs stretching like fili-gree hammocks from plant to plant, wandering slugs, black and glistening, and a pungent, resinous tang from the adjoin-ing firwood mingling with that of overturned soil. Sea and sky were a misty blue, heralding heat, while a glimpse through the trees along the island coast showed that the tide was out, leaving a wide expanse of sand, wet and rippled where the grains had been alternately compressed and scat-tered by the retreating waves. Unhappiness and argument seemed a desecration to such surroundings, but every time the Boss chuffed past me up a parallel row, he had some word of criticism for my work to make me fester, while Jack flashed at me a bitter, accusing eye.

As the sun rose higher, the air began to quiver with heat and the shrilling of cicadas. Digging in the dry hard ground sent beads of perspiration trickling down my back. The Doctor deserted us to fetch beer from the house. Jack dragged his shirt over his head and flung it onto a stake where it hung like a flag.

The beer and heat acted as a thaw. While the Boss reclined under a tree smoking, Jack and I went down to the beach for a swim. The sea was warm and the dry sand burning. We stayed too long, briefly happy to float and dive and splash like children, to watch the dogs chasing gulls, the fish leaping, gannets swooping and to marvel at the green limpidity of the water. But we were hotter than ever when we had climbed back up the hill to find the Doctor snapping with impatience.

The long morning had tired us all and by afternoon Jack and I worked unwillingly, backs scorched and aching under the high sun. We watched the Boss driving himself relentlessly up and down the clearing, doggedly determined to see the job completed before nightfall. He laboured slowly and meticulously, pressing the earth around every seedling with loving care, then standing back to judge if the plant was straight and firm. But Jack tossed each young tree into its hole and kicked back the soil with his boot, unconcerned, in his mood of despair, whether it would live or die. He had no money and no plans and the cheerful, self-confident optimism, which had once lit up his character, seemed to have deserted him. Towards four o'clock he suddenly threw his spade on the ground. He had had enough, and disappeared down the path into the firwood out of sight.

A little later I noticed a thin column of blue smoke rising from the shore. He was signalling for Ben with a bonfire of seaweed on the beach. But Ben did not come. Perhaps he had not returned from fishing, or was spending the day in town. When I went down to the steading an hour later to milk the cow and get tea, I found Jack still waiting with his swag beside a smouldering heap of kelp.

He strode across the shingle. 'Why the hell do you have to live on an island? I can't get away from the place now.'

I said I would ask the Boss if I could borrow his boat and take Jack to the mainland. But the Doctor would not hear of such a plan. Jack was in disgrace and must apologise first. Besides, my help with the tree planting was needed more urgently than ever now that one worker had gone on strike.

Poor Jack stood on the beach hurling pebbles into the water with vicious force in his anger and frustration, while indoors the Doctor struggled to eat his tea without choking. My patience had expired. I was on nobody's side, equally exasperated with both men.

Then glancing out of the window as I rose to fetch more hot water for the teapot, I was arrested by the sight of a boat in the bay. For a moment I failed to realise what had happened and then it dawned on me that the boat was the Doctor's dinghy and the man rowing it was Jack. He had almost reached the farthest arm of the reef.

I turned round to tell the Boss, my expression not quite as solemn and horrified as was appropriate. 'Jack's taken your boat.'

'What!' He ground out a half-smoked cigarette on his plate and rushed to the window. His face became distorted with fury. He was hopping mad. I watched him chase down the beach, shouting obscenities, dancing on the pebbles like someone demented. Safely surrounded by water, Jack rested his oars and waved, before resuming the long pull to the mainland.

The weekend had proved disastrous for us all. The Doctor had not found his 'Happy Isles' while Jack, through his own uncurbed tongue and temper, was now banned from returning. Aroa was to be forbidden territory for him from now on. As for me, I saddled a horse next day and set off to muster my sheep for the shearing in a rebellious mood. I had had enough of company.

Rounding up the stock single-handed was never easy, especially the ewes with young lambs. I had to start out at dawn before the heat of the sun made the stock hide away in the shade of the bush, from which they were always reluctant to move. I could never have managed without Jock and Glen, who ran to and fro behind the stragglers with tireless energy. Skipper, the all-purpose horse, was equally invaluable. He was sure-footed on the precipitous cliff-paths, fearless of cattle, nimble as a polo pony, yet willing to stand and wait patiently with reins hanging loose if I had to dismount and hunt out sheep from the rocky shore or dangerous cliff-faces.

If I was mustering cattle I used a stockwhip, for the gun-shot cracks of the flax tail lashing back on itself helped to bring the malingerers out of the trees, and barking and yelling all helped the business of droving. Sheep needed more subtle methods. Making a noise or rushing headlong after them would only have caused panic. They had to set their own pace and be allowed to file along their familiar paths, following one another like tributaries running downhill to join the mainstream, with a few wise old matriarchs leading the way.

The dogs had developed into a good team, Glen creeping and stalking silently round the perimeter of the flock in a wide arc, while Jock took charge at close quarters, darting behind their heels, urging them forward with eager energy, and barking to order if a sheep loitered to graze. The chief snag to sheep-handling was not getting the flock to the pens. Arduous as this could be on so large and rugged an area, it was afterwards, waiting for help, which was wearing, knowing that the arrival of any extra hands depended entirely on the weather. If the sea was too rough, no one could come.

The ten-acre paddock could hold the waiting sheep for a day or two, but if a longer delay occurred I was forced to release them back into a larger enclosure where they would not starve. Again and again, when arrangements had been made for any movement of stock, for buyers or vet or casual Maori labour to come to the island, the wind sprang up overnight, the sea became impassable for days on end and my plans had to be shelved, or changed. Earlier in the year, I had struggled to castrate and tail most of the lambs on my own before help arrived and now, in frustration, I embarked on the shearing, while the usual gale blew.

At first nothing went right. The belt on the diesel-engine snapped, and when I had mended it I found that I had made it too tight. Some crucial part from the innards of my shearing hand-piece fell out and then, just when I had everything all set to go, it began to rain, and only a handful of sheep already penned in the covered woolshed could be shorn.

Five days after the muster, the shearers arrived. I watched them flailing off the wool with effortless ease and speed, curling each sheep round their moccasined feet for the long blows from tail to head. The New Zealand style of shearing looked deceptively simple, but I knew half the art was in handling the animal, and in having sharp combs and cutters. When each sheep was finished, it was booted out of the shed through a pop-hole door to leap joyfully in the air as it escaped to freedom. Meanwhile, the fleece was gathered up and tossed out like a mat, so that it landed on the slatted table spread flat and ready for sorting and rolling. It was my job to consign the various parts of the fleece to their individual piles – locks, bellies, topknots, dags – and then to roll up the main body of the fleece and put it into the bag on the wool-press. I had to keep the sheep pens full, catch ewes for the shearers, and be ready with antiseptic for any cuts. I also had to brush the shearing-board, and rush off from time to time to make pots of tea and hurried meals.

After three days' work, when all the sheep were shorn and the men had gone, I set out to look for strays. There were bound to be a few which had evaded the muster. These I clipped myself. Then I finished packing the wool-bags, sewed them up, and waited for Ben to collect them, three at a time, for transport to the mainland in his boat. It was essential to keep the wool dry, so help was needed to load and unload the bags and take them up the beach on Ben's horse-drawn boat-sledge to the car-shed, where they would await a lorry.

Work in full spate, followed by lulls of idleness, was typical of my life at Aroa. Either I was punishing myself to complete some marathon task, or basking in the sun and the sea without a thought for the day of the week, or the time. But the calm which I loved and valued so much was broken intermittently by a new disturbing element.

Telegrams, like sudden claps of thunder, began to arrive with increasing frenzy from Jack. Each time a wire was received at the post office on the mainland, the poor school-cum-postmaster had to send a messenger down to the shore to find a fisherman willing to bring the message over to me. I became more and more incensed, not only at the futile bother this was causing other people, but at the nature of the telegrams. Marriage proposals and protestations of love were followed by demands, commands and threats to name the day, to waste no time, to marry, or get off the island. Then bills for repairs to my van appeared. I had lent the vehicle to Jack after his last visit to the island, feeling sorry for him as he was out of work. He had borrowed it to drive himself down the length of New Zealand to the Southern Alps, where he had hoped to get a job deer culling. Now

135

this new, uncharacteristic behaviour worried me. It was so unlike him. Where was his sense of humour, his open, generous, warm-hearted nature? I felt both uneasy and hurt by his change of attitude.

Meanwhile the Doctor was busy inventing a raft for transporting stock to the island, and planning how we should construct it on the shore when he returned to Aroa with a party of friends at Christmas. The basic idea was very simple: a platform of railway sleepers laid on a flotilla of empty water-tight oil drums. Metal hoops round the drums would attach them to the sleepers. Then, on top of the platform, a horse- or cattle-box could be bolted down when required. Propulsion could be by two outboard motors attached to the stern, or possibly the barge could be towed by Ben's boat. It sounded an ingenious idea, but there were two major snags. No one had ever seen a raft like this in use, so we could not tell if it would work. And it would be so heavy and awkward to move that it could only be constructed on the shore between tides, and floated off when the water rose. This meant a strict time-limit on the building process, and disaster if the tide beat us.

Having heard about these plans by letter, I awaited the construction of this strange Kon-Tiki with interest and a few misgivings. Knowing the traumas which always seemed to surround the Boss and his frantic working holidays at Aroa, it appeared too much to hope that this Christmas would prove to be a continuous and miraculous season of merriment and goodwill.

In fact, I spent Christmas day happily picnicking on a mainland beach in bright sunshine with the schoolmaster, his wife and their baby daughter. Their habitual kindness to me was a great source of pleasure and our easy, relaxed day on the shore was, in retrospect, like the calm before the storm.

*   *   *

The Doctor and his guests had not been on the island two days when, late one night, while everyone was sitting about among a debris of empty coffee cups and glasses, Jack burst into the house looking wild and dishevelled. He held a knife in his hand, the sheath-knife which had been used so often for gutting fish, skinning sheep, or slicing bread and peeling apples on summer expeditions. In horror I watched him threaten the Boss with his weapon before two of the men leapt to their feet and, grabbing his arms, restrained him. The danger was over in a matter of seconds, but the aftermath of fear, fury, indignation and consultations on what to do about this unmanageable young man went on until it was nearly dawn. Jack was overwrought and desperate. He had abandoned his job in the south, he had no money, nowhere to live and looked half-starved. He had taken without permission a Maori fishing boat to reach Aroa, no doubt expecting to find me alone and perhaps hoping for refuge. Instead he found the island milling with people and, in a moment of panic, he had lunged at the man he now detested.

I felt sick with misery as if somehow I was responsible for everything without knowing how, or why. The changes in Jack, his threatening behaviour, the distress he was causing everyone, must all be my fault. But what could I do? Surely no one expected me to marry to save him from bankruptcy and despair, and myself from being molested and persecuted. My affection had been genuine in the past, but the more Jack tried to force himself upon me, the less I wanted him and the more I valued my freedom. He was too insistent, too anxious to dominate, too intolerant. What use was burning passion, if it scorched and hurt, destroying the trust which is the essence of love? Such a marriage could only have proved disastrous. So I made my unwillingness clear yet again and retired to bed, bolting my door, while the men shepherded Jack to the bunkhouse to snatch what rest they could until it was light.

137

In the morning the Doctor took command of the situation. He had made up his mind. Jack was transported to the mainland and handed over to the police. He was charged with trespass, threatening behaviour and conversion of the fishing boat. He pleaded guilty and was put on a year's probation. It was a sad business and when I saw my van where he had abandoned it on the edge of the beach with two flat tyres, an empty petrol tank and every sign of having been abused and misused, I could have wept with bewildered disappointment.

Back on the island, we resumed work on the raft in subdued spirits. We had painted the oil drums and metal hoops to save them from rusting in the salt water, and now all we needed was calm weather to assemble the vessel and launch it at high tide. Eventually a promising morning dawned, so the Boss and I and our other helpers spent the next five hours bolting the raft together, panicking to get finished as the sea crept closer and closer up the beach. Work was completed just in time and cheers went up as the Kon-Tiki finally began to float.

However, the maiden voyage was not a success. The outboard engines fixed on to the raft were not sufficiently powerful and there was no means of steering the vessel. Towing with the Doctor's dinghy was not practical either. Obviously Ben's heavier fishing boat would be needed, so the raft was anchored temporarily in the bay and a smoke signal sent up for Ben.

The first victim to be transported on the raft was a horse which the Doctor had purchased locally, so the raft was towed across the straits the following day and beached on the ebb tide. Then all hands helped to bolt the patent horse-box on to the railway sleepers and construct a ramp to encourage the alarmed horse to ascend the gangway. The mare was very frightened and reluctant to cooperate, but slowly, inch by inch, we pushed and pulled and wheedled

138

her forward. Finally, the door was wedged behind her and all we could see were her ears twitching over the partition, while we listened anxiously to sounds of kicking and splintering wood.

At last the sea began to lap around the raft and lift it from its bed of sand. The mare snorted and plunged in her box as she felt herself being tipped from side to side. Ben had launched his boat and was taking the strain of the raft on a tow-rope. A small audience of excited Maori children watched the strange barge bob over the first few breaking waves into the quieter water of the bay. I travelled on the raft, trying to reassure the terrified horse and listening with growing alarm to her kicking the box to pieces. Already several planks were cracked. It was a nerve-racking voyage. Once away from the shelter of the mainland coast, we headed into a strong cross-current of short choppy waves, and the drag of the water slowed down progress.

We beached the raft on Aroa at high tide, unbolted the door of the battered horse-box and coaxed out the mare. One of her legs was badly gashed and she was never fit to ride again. Perhaps the Doctor should have listened to Ben. Long ago, when Maoris farmed the island, horses were taken to and from the mainland by towing them behind a boat, giving them no choice but to swim. And we later ferried several across like this successfully. Horses are strong swimmers, but three miles would be a dangerous marathon unless the sea was perfectly calm.

CHAPTER TEN

The New Year brought drought. Each day a fierce sun shone in a vivid sky, yellowing the pastures and cracking the earth. The sand-dunes were too hot to lie upon and every path was powdered with dry dust. After grazing the gullies in the early morning, the stock hid in the shade of the bush until evening. The dwindling springs were poached and polluted, and the sapless grass rustled like straw. In the garden, the soil had set rock hard and the sun had scorched the lawn brown. Every door, every window to the house stood open, day and night. By ten o'clock it was too hot to work. Then the island lay sleepy and deserted, curing in the sunshine while the air and water shimmered with heat and light.

Every morning I woke to the indolent slapping of waves on the shore and stayed listening until I could no longer resist the invitation to swim. Floating, I could watch the sheep filing down from the hills to graze and the seabirds flocking to their fishing grounds. My existence was not so different from theirs. Most days I went out in search of food, gathering shellfish, digging taro roots, picking grapes and peaches, drifting in the dinghy off the reef waiting for shoals of snapper. I would milk the cow and hunt for eggs. There was no hurry, no urgency. Tomorrow would do. I was becoming like a Maori in my attitude to life, as well as in my appearance.

Visits to the mainland became less and less frequent. I sold the redundant van and bought another eighty shearling ewes with the proceeds. My first wool cheque arrived and it was satisfying to hear that the island fleeces had fetched top price at the Auckland wool sales. I felt as if I was riding on the crest of one of those great, rolling, curling ocean waves which reared up in the sun shot with green translucent light. Not only was my ambition to be a sheepfarmer fulfilled, but here, all around me, was the landscape I loved and needed.

One hot afternoon in late February I had been baking bread. The new loaves, warm and crusty, were cooling on a rack on the table. Though the doors and windows were thrown wide open, the smell of bread filled the air. The dogs panted in the shadowed porch, exhausted by the heat. My back was turned towards them while I washed a mixing-bowl at the sink. I heard a scuffle of movement, a single bark and swung round, startled. Jack! He had caught me completely by surprise. He stood in the doorway staring at me, his clothes torn and wet, his chin dark with stubble. I could tell at once he was desperate.

After abandoning an attempt to reach the island in a rubber dinghy, he had taken a Maori fishing boat again and landed unseen, behind a headland. He was out of work, without money and had not eaten for several days. If I would not marry him, he did not want to live, he said.

I persuaded him to sit down with a drink. Then I made him eat and tried to comfort him. But however much pity I felt, I could only continue to refuse him. He was destroying my affection and trust.

I wanted him to go back to the mainland quickly and return the boat before it was missed, but he would not listen to my pleading and reasoning. His threats frightened me. He spoke for the first time of suicide. He seemed so subdued, so defeated, as if tired out in body and soul, with nothing to tell me, nothing to ask.

I decided that he was suffering from physical exhaustion and eventually persuaded him to lie down and rest. He slept and slept, right through the evening and night. I began to feel happier, convinced he would be different when he woke. The bright light of morning would cheer him. The ordinary down-to-earth routine of milking, feeding the poultry, cooking breakfast, would bring him back to reality. Surely no one could think seriously of suicide before lunch! I would help him decide on a plan of action: somewhere to go, a new job, and encourage him to leave the island before any Maoris came searching for their boat.

Early next day I went out to milk the cow before there was any sign of Jack. It was a lovely morning: a blue sky, no wind, and all the promise of heat. The water in the bay, still shadowed by the headland, was a deep clear green, and from the paddock I watched white wide-winged gannets poise, plummet and pierce the sea's calm surface, their swift impact flinging up fountains of spray. When I returned to the house, I found Jack in the kitchen with the Doctor's rifle in his hands. He was in a frantic, fiendish mood and I was petrified.

He began to take pot shots at seagulls through the open window. The sound and smell of firing made my stomach feel strange. Trying to keep calm, I suggested that he put away the gun, and that we should talk things over. He refused. There was only one thing he wanted and I could not give it. He began to threaten me. I pleaded and reasoned and even attempted to make him laugh, but words were no use.

Turning my back, I walked slowly and deliberately towards the door, thinking that if it was me he wanted to shoot, now was his chance. Nothing happened. Once outside, where he couldn't see me, my pace quickened, I wanted to run, suddenly panic-stricken. Afraid of venturing into the open now, of crossing the paddock, or even the lawn, I slipped into the toolshed and immediately regretted it, seeing the horrible weapons around me, the axes and hammers and billhooks. I had to get out of there, but stepping on to the path outside, I saw Jack coming across the grass.

He was unarmed, but frantic with frustration and catching hold of me, he pushed me against the wall. He hit me several times, but I fended off the blows half-heartedly. I did not want to fight. The brilliant sunshine, the sheep filing quietly across the yellow pastures to find shade, the sea, vivid, indolent and sparkling beyond the flat-leafed fig trees, made our shouts and blows seem small and ludicrous. It was like a brawl in a church, a desecration of serenity. I begged him to wait, to think, to do nothing in fury which he might later regret, but he would not listen to me. He was ruled by a single obsession.

I told him that I would rather die. As he looked at me, I could tell that he was trying to measure my strength, wondering if I would finally give in, maddened by my conviction. Then swinging round, he started towards the house to fetch the gun, while I stood there in the garden, leaning against the sun-warmed planking of the toolshed wall. It was a curiously calm moment, now that I knew what to expect. My decision was the only possible one: clear, logical and sure. I loved someone else, not Jack, and

144

therefore I would not be possessed by him. I set too high a value on loving.

Dreaming, I must have let nearly a minute slip by with Jack out of sight, before I recognised my one chance to escape. A Maori boy had already been over to Aroa to collect the missing boat, but the Doctor's dinghy sat on the shingle, half in and half out of the water where Jack had dragged it, thinking to leave the island, and then changing his mind. Now I fled across the lawn, dodged between the flax bushes bordering the shore, and ran down the beach. The pebbles rattled under my feet and I was afraid he would hear me. In frenzied haste I pushed the dinghy into the sea, and hauled up the anchor from the shallows. Somehow in my hurry, the rope became tangled and I floundered knee-deep in water, tugging in silent panic. And then I heard him running over the shingle.

My back was to the beach. The rope came free at last and I was tossing the final loop into the boat and preparing to jump aboard when I knew by the splash of feet that he was behind me. Before I had a chance to turn and see what was coming, something cracked down on my head with terrifying force. I fell, and, as I fell, he hit me again and again until the butt of the rifle broke off and dropped on to the pebbles beside my face. I had dragged myself a little way and was now lying half out of the water. I could feel the blood running down my neck and round my ears. There was a long, taut silence, when I knew he was standing over me, looking at me, and I tried not to move a muscle so that he would think I was dead. Then I heard the crunching of shingle as he walked away.

Paralysed with shock, hardly conscious, I waited and waited, listening until there had been no sound other than the lapping of water along the shore for a long time. Eventually, lifting my head very cautiously, I looked around to find the beach deserted, but any feelings of relief quickly evaporated when I saw over my shoulder the boat, the only boat, the one way to safety, drifting out to sea.

Desperation rather than strength drove me to my feet and into the water. There came a point when having swum about two-thirds of the way to the dinghy, I realised that unless I caught up with it, I should never regain dry land. It could only have been will-power which kept me afloat and heaved me, gasping like a fish, over the gunwale into the salt pool of water amidships. Even then I did not feel safe, but imagined Jack to be in the house and expected any moment that he would see me and shoot from the window across the bay. Bending low, I struggled to start the outboard motor, but without success, and so picking up the oars, I began to row.

Blood kept running down my face into my eyes, splashing on my clothes and into the water swilling at the bottom of the boat. I was feeling dizzy now, and weak. After battling for about ten minutes, I realised the futility of trying to reach the mainland by rowing. My strength was ebbing too fast and so I made another attempt to start the engine. This time it responded. Drooping over the tiller, splashing my face intermittently with the cold sea water to keep conscious, crying with the pain and horror of what had happened and watching almost fascinated, the dinghy filling, so it seemed, with blood, I steered for the mainland. Three miles away the narrow half-moon bay and the white shacks of the fishing settlement were visible in the bright morning sunlight and the open face of the sea shone bland and blue.

CHAPTER ELEVEN

A Maori woman came running down the beach and threw a towel over my head and a fisherman dragged the boat on to the sand. Then they fetched the schoolmaster, who was the only person with a car, and he drove me to hospital. A small country hospital, thirty miles from Aroa.

I stayed there for three weeks. A fracture of the skull was suspected but never proved, and a dozen stitches were all the medical aid needed. My survival was a miracle, the doctors said, and thanks solely to my having a head like a crash helmet.

After I had spent twenty-four hours in a darkened room, a policeman was allowed to come in to take a statement, and I heard how Jack had been arrested. An armed party had gone over to the island in two of the Maori fishing boats and surrounded the house. Jack, the exhibitionist, was acting the last scene of his melodrama, sitting in the living room with the broken gun beside him, a note written for posterity on the table and a bottle of my home-brewed wine at his elbow. He accepted the handcuffs without resistance. Apparently he believed I was dead, and it was not until the following day when he was charged with assault, not murder, that he knew otherwise.

There were headlines in the newspapers and a whispering and peeping of nurses on the verandah outside my room. In the evening the Boss arrived. He had flown to the nearest airstrip and greeted me with a large bouquet of flowers. He was deeply shocked by what had happened and full of concern for my injuries, but he spoke heatedly of Jack. It was dangerous ground for conversation. It made me want to cry. I was glad when we talked instead of Aroa and he wrote out a list of all the things which I wanted him to do next day. Check the cow's udder (fortunately she was due to calve soon and was being dried off), give the poultry a generous supply of corn, bring the dogs and cat to the mainland and find my dressing gown and various belongings which I needed at the hospital. While he scribbled, I suddenly wondered how much he cursed ever having met me. He had had a bee in his bonnet about some harm coming to me, imagining accidents, illness and molesting fishermen. I smiled to remember him fitting bolts on the doors, turning the house into a fortress. He had been concerned and kind. But now he stayed too long. I was tired out and horribly aware of the hammers in my head before the hospital doctor came and urged him to leave me to sleep.

Until this misfortune, I did not know how many friends I possessed. People I hardly knew came to see me, both whites and Maoris. Bob walked ten miles to visit me one afternoon, then returned to his yacht to sail out to Aroa and look at the sheep and cattle for me. The schoolmaster and his wife came regularly with gifts and comfort. Each day brought letters, flowers, books and visitors. Even my friends in Taranaki had not forgotten me. I was spoilt. But at night

148

the thought of so much sympathy made me weep. Jack was the one in need of pity.

I could forgive his action. It was a momentary, never-intended crime, but it made me toss and turn to think how he would have to pay for it. He might be middle-aged before he was free again. Then, one day, someone told me that just down the hill in the township Jack was to be seen mowing the lawn at the police station, playing with the constable's children and eating ice cream cones in the garden. So I felt less sorry for him and occasionally, lying awake at night, I was frightened at the possibility that he could so easily escape from the small local gaol and come up to the hospital after me. I had trusted him implicitly once, but that was over.

My thoughts churned over: worrying about Jack, worrying about the stock on the island, wondering whether they had enough grass, enough water, and if they were straying down the cliffs for the sake of a green bite. When people asked if I would go back to Aroa, I was bewildered. It had never occurred to me to consider any other plan; I could not wait to get home. If only the nurses had not taken away my clothes, I believe I would have crept out one night and started walking to the coast. I was sick of my hospital bed, the heat, the noise, and the lukewarm cup of tea thrust into my hand at dawn. There was a seat outside my door on the verandah where a patient from the men's ward sometimes sat. He was old and frail, hardly alive, hardly sane. It depressed me more than anything to see him groping for the bench, groping to understand the bright chatter and teasing of the nurses. My room was like a florist's shop, but no one took flowers to him. Then there was a little Maori boy who had injured his arm so severely on a moving belt while helping in his father's milking shed that it had had to be amputated. He could not have been more than eight, a quiet black-eyed child. I never heard him cry. It humbled

149

me. There was so much courage and kindness around me, yet I longed to get away. I was desperately homesick.

On my return to the island, I found the pastures scorched and tinder dry. The earth had cracked like crazy paving and only two springs flowed. During the few weeks I had been away, the stock had become visibly thinner. The rams were running with the ewes now, and should have been flushed on good grass to bring a higher percentage of lambs in the spring instead of having to hunt for every green leaf deep in the gullies and under the canopy of ferns and manuka. Fallen fruit lay rotting beneath the peach trees and on the hillside vines, grapes were drying like raisins in the sun. The figs had ripened, their skins a dark, shiny purple-brown, and the cow had broken into the vegetable garden, ravenously eating every carrot, cabbage and lettuce which she could find, leaving the plot as barren as a desert. A multitude of spiders had been busy in the house, running up net curtains, and the cat, after her holiday on the mainland, was developing a suspicious abdominal bulge.

Two days later the rain came, and with it the wind. It seemed like the end of summer, for the air cooled rapidly, but the rain was manna and the stormy sea a blessing. Since coming back to Aroa, I had been more nervous than I cared to admit to myself. If the dogs barked suddenly, I jumped. If I heard a strange sound, or a far-off boat, I grew tense and stopped whatever I was doing to strain my ears and wonder. But a rough sea meant safety: no one could reach the island.

I expected to be asked to attend the court case to give evidence but no one informed me when or where it took place. Then friends sent me press cuttings. I was appalled at the coverage. Not only had Jack asserted sex with me had taken place, but doubts were also thrown on my relationship with the Doctor! I was described as his housekeeper and paid employee. A far cry from a share-farming business partner making an independent living from my own flock of sheep.

I resolved to put the record straight and wrote out a statement, even offering to undergo a medical examination, if necessary, to prove that I had never had sex with Jack or anyone else. My plan was to attend the court when sentencing took place and make sure the truth was heard.

Thankfully the sea was calm on the appropriate day and friends from Taranaki, who had known Jack, drove up and took me to the courthouse in Whangarei. I handed in my statement and soon afterwards a probation officer invited me into his office. He had spoken to Jack, who had withdrawn his assertion. The officer surveyed me patronisingly and reproved me as if I was a child, not twenty-five.

'Your parents won't like this, will they?' he said. 'You be a good girl in future.'

I glared at him and left the room. Didn't he realise that it was precisely because I had been 'good' that Jack had tried to kill me? Jack was sentenced to three years in gaol for assault causing grievous bodily harm. He looked subdued and somehow smaller.

Later, as I left the courthouse, I saw him being escorted down the street with a large police officer on either side. He spotted me, paused, appeared to plead with the men, who then crossed the road with him to where I stood hesitating. They stepped aside briefly for a few moments while he spoke to me.

'I'm sorry,' he said. 'I love you.'

Then he was marched away. That was the last time I saw him.

The grass began to grow, and within a few days the hills had changed from a dry rustling yellow to a rich leaf-green. Enormous mushrooms, like white inverted saucers, sprang up in the damp gullies overnight. The dust was washed

151

away in the rushing streams, staining the sea brown where the fresh water flowed into the salt. Sometimes, when I sat in the house after dusk, there blew such squalls they threatened to burst the windows and lift the roof, and I found myself clutching the arms of my chair as if I expected to be swept away.

There were days of torrential rain. The paths on the island became streams. The garden was flooded. Water seeped under the doors into the house. A blue mould grew in the cupboards, and clusters of fungi sprouted between the floorboards in the disused bedrooms. Even the roar of the surf was overpowered by the drumming of rain on the corrugated roof. The drought and then this spate of water caused several landslides, and the sea was thick and opaque with island soil. For nearly a week the hills of the mainland hid behind heavy cloud and a curtain of sea mist, but I rejoiced. There was peace and security and a wild beauty everywhere. And there was grass. I was busy each day, gathering the ewes to the rams, picking watercress and giant mushrooms, combing the debris on the stormed beaches, hunting for wounded birds and rare shells.

It was only when my head began to hurt, as it did sometimes, or when I thought of Jack that I was miserable. We were both captives, for I suppose the island was a form of self-imposed prison. But whereas he would wait dreaming and stamping for release, I could not think, or care, beyond this present time. I wanted nothing but to love and value Aroa, to get drunk alone on a strong, salt, sea-green wine.

Solitary confinement was no hardship to me. I had grown accustomed to it years ago in the Cheviot hills and on Exmoor. Besides, I did not consider Aroa solitary. Everywhere one looked there was life, wader-birds strutting along the shore, fish swimming in shoals, cows munching kelp, ewes cropping the banks of the creek, pheasants calling from the

152

fir plantation, sand worms burrowing, plants growing. And wherever I went the dogs came with me, eager, alert and exuberant, ready to fly round the sheep at a signal, or chase along the sands like children, or fling themselves down beside me, tails thumping with unrestrained affection. They were my shadows.

One morning I rode down the track to the steading and saw an unknown man sitting on the beach beside a small suitcase. After a comparatively quiet day the wind was rising, whipping up white froth on the short, steep waves in the channel. The Maori fishing fleet had returned early and already the sea was too rough for my boat. Whoever had come to the island would have to stay.

My reception was hostile. 'Who are you?' I demanded. 'And what do you want?'

I confronted a man of about thirty: slim, fair-haired, pale, reasonably dressed and polite in manner. He explained that he was interested in ornithology and that he had persuaded a Maori fisherman to bring him over to the island, hoping that he might camp for a night or two. When I pointed out that he appeared to have no tent, he looked uncomfortable. 'And what about food?' I said. Had he thought of that? He could not live on fresh air.

He became more and more apologetic. No, he had no food, but perhaps he could catch a fish. I half expected him to produce a piece of string and a bent pin from his pocket, but instead he stood there looking dejected and helpless.

'Well, I can't take you back, the sea's too rough,' I pointed out. 'And the Maoris won't come over now. It's getting dark. You're marooned.'

This information aroused the glimmer of a smile, but it was quickly extinguished as I marched him round to the bunkhouse and handed him a couple of blankets and a candle. Half an hour later, I offered him a very burnt, unpalatable supper. Not that I had burnt it deliberately.

153

It was just a common flaw to my cooking, especially when I was flustered. Our conversation was stilted and embarrassed. My visitor informed me that he was a poet and his name was Milton. That seemed so very odd that I took care to bolt my door when I retired to bed. I slept uneasily. Who was this stranger and why had he come? His arrival was an unfathomable riddle.

A full gale blew in the morning and there was no hope of crossing to the mainland.

'You'll have to work for your living,' I said to the stranger.

'Oh yes,' he responded eagerly, 'I will. Anything at all!'

'Can you kill a sheep?'

He blanched. No, that was something he definitely could not do. I explained that there was a ewe dying from a gangrenous prolapse and that we must destroy her to save her suffering any more pain. She might go on living in agony for another week. It would be cruel to keep her alive.

He watched me sharpen my sheath-knife with an expression of incredulous horror on his face, while I felt agreeably relieved at this proof that he was not a natural butcher. As we walked up the paddock to the place where the ewe lay straining, I coaxed and threatened, jeered and begged him to perform the deed of mercy, but he was adamant. He would do anything I asked, except that.

'All right,' I said contemptuously. 'I'll cut her throat and you can do the skinning and burying.' I was so angry with him and so anxious to see the sheep put out of her misery that I slashed without hesitation. Of course he had no idea how to skin a carcass anyway, but at least he tried and we were so busy that it was not until the job was completed that I noticed how my companion and his clothes were splattered all over with blood. When we got back to the house I lent him some overalls, which proved very tight and short in the leg. He began to laugh. He felt such an idiot, he confessed, as he scrubbed his trousers under the cold water tap.

I allowed him no respite. In the afternoon, we took a horse and sledge and saws round the bay to a large log washed up on the beach. There was enough wood for many a winter fire. We sawed for three hours until finally, in desperation, Milton showed me his hands. They were raw with broken blisters. Back at the house, I bandaged him up. Perhaps, after all, he was a poet. He was not a manual worker, for certain.

The following day the storm continued to lash the island and beat up grey menacing waves in the channel. The lost, inaccessible mainland lurked behind a shroud of tossed spray. I did not know what to do with my unwanted visitor. I put him on a horse and he promptly fell off, adding a limp to his other injuries. He seemed totally impractical. In the end, I left him to his own devices, so he sat about the house reading books and writing fragments of verse about wind-thrashed islands, sheep bones and tyrannical women.

I suggested we ate the best cut off the ewe, but Milton professed to be a vegetarian, and munched his way through sheaves of watercress and slices of brown bread with cream cheese. He never asked to borrow my binoculars or exhibited the slightest interest in the bird life of the island. Nor did he betray any eagerness to depart, but examined the wild impassible sea each morning with something suspiciously like relish. We both thought each other very odd, but gradually we adjusted to our enforced companionship. We began to relax and while he teased me for my eccentricity and barbaric way of life, I laughed at him openly for his soft useless hands and wooden-leg gait and all those weaknesses, which made him such a singularly hopeless castaway.

On the fifth morning, the sea had calmed to an easy swell and I announced to my visitor that I would take him back to the mainland. He looked woebegone. He did not want to go. The island had cast its spell.

He begged to stay a little longer, but I had made up my mind. Spongers must go, especially spongers who were beginning to get round me with little offerings of chopped kindling, trimmed lamps and scraps of verse laid on my breakfast plate.

A week later I found a thank-you letter from Milton in the mailbag. It was a good letter, articulate and full of wry jokes about the injuries he had sustained and the unforgettable experiences he had survived. One day he would come back, he said, and ask me to marry him. My astonishment was capped a few days afterwards when gossip reached me that my strange uninvited guest had been an escaped convict. What possible crime could a man have committed, who refused to kill a sheep?

My second lambing at Aroa was much harder than the first. Not only had I more sheep to manage single-handed, but they were in poorer condition, due perhaps to the prolonged drought and then torrential storms during their pregnancy. After one particularly vicious gale, I found not merely birds and fish as battered, stranded victims, but six ewes that had died overnight. The elements had simply been too harsh to withstand. What chance had weakly new-born lambs when the weather turned hostile? I worked to the point of exhaustion to salvage all the lambs I could, but sometimes it seemed a losing fight. I ran out of milk for those in need of bottles, and pens for those in need of shelter. When I went into the paddock to milk the cow, a flock of tame lambs followed me like children behind the Pied Piper. The ewes were too thin to lactate well, and though the best of the single lambs thrived, many of the twins became stunted.

The cattle fared better on the far side of the island, less affected by the build-up of intestinal worms, which could play havoc with the health of sheep. Calving too was seldom any trouble, and the milk supply was ample for the single-suckled Herefords. Castration of the bull calves was the only really difficult challenge. I preferred to tackle the calves while very young and easy to hold, but then their mothers tended to be fiercely possessive, and more than once I had to take rapid evasive action, dodging round the clumps of flax, manuka and cabbage trees which encroached upon the open pasture.

It was not an easy spring for me, and just as I was beginning to relax towards the end of the lambing, finding time to fish and swim and sail again, I was suddenly banished from the island by a totally unexpected event. Looking out of the window one morning, I saw a boat heading for Aroa. No one was due to come and I wondered uneasily who was on their way. Through binoculars I counted three figures.

As they approached, I recognised Ben, another Maori, and the local schoolmaster. I went down the beach to greet them. They had a gun in the boat and at first I imagined they must be coming to shoot wild turkeys, or perhaps to beg a sheep for a Maori tangi, but their faces were too sober and I waited apprehensively for their news.

They brought a message from the police that I must leave the island at once. Jack had escaped from prison. Very generously the schoolmaster and his wife invited me to stay with them, but they were under instructions not to allow me to venture anywhere alone, or return to Aroa without an armed escort. Apparently Jack had slipped away from a working party in a field and was believed to be heading north. It was thought that he carried a knife. I listened to this news with horror and indignation. How like him to try to escape! He had too rebellious a nature to accept the discipline of prison for long without making a bid for freedom. But to come back to the island, to come back for me! The possibility made my blood run cold. Surely he knew

157

better than to leap into the same trap. Surely he realised that the police would be waiting for him.

Angrily I packed a few belongings, gave the hens enough grain to last them several days, released the calf with the house cow, and collected up the cat, dogs and four bottle lambs, which I had no option but to take with me to the mainland.

The first few days of exile were spent in a state of mutinous fear. Jack was as likely to find me at the schoolhouse as anywhere, for this was the last staging post to the island, and a place where I frequently stayed. It put my friends in danger too and I lived in a turmoil of outraged fury at the unfairness to them, but I did not know where else to go with all my dogs and lambs. Then, from the farming point of view, things could hardly have been thrown out of gear at a worse time. The lambing was not yet over, and calves had started to arrive. The only solution was to go over to the island whenever I could and try to catch up on the work waiting to be done. As usual Ben came to my aid, and took me across the water in his boat, except on stormy days, and fished off shore while I hurried from job to job, easing the surplus milk from Blacky's udder, inspecting new lambs and calves, tailing, ear-marking, and checking water-holes. And each time Ben brought his loaded gun, but gradually, as the first tension wore off, it became a joke between us not to forget our weapon.

Only at night did my fear run out of control. I could not sleep for imagining strange sounds, footsteps in the garden, the rattle of a window, the banging of a door. I rehearsed all the things which I should say to Jack: how I would appeal to his reason, his morals, his religion, his love; but I knew his violence too well. Supposing he would not stop to listen? Sometimes I felt sick with terror when I thought of him heading north, perhaps daily coming nearer and nearer the island.

After a fortnight there was still no news of anyone having seen Jack, and the police agreed to let me return to Aroa. Two friends offered to keep me company for a few days, so we all crossed over to the island in Ben's boat laughing and joking while we hung on to the cat, lambs, dogs and gun.

It was good to be home and my fears were banished by the pleasure and excitement of returning. Even when the boys suggested that Jack might be hiding on the island, I was not worried. This was the one place I wanted to be, and if I was going to get a knife plunged in my back, then I would rather it was here.

That first night one of the boys rushed into my room with the gun. He had been woken by steps approaching, rattling the pebbles along the beach. We listened intently. It sounded unnerving, but I guessed what it was and presently we were both chuckling with relief at the reassuring chumping and blowing of a horse munching seaweed under my window.

After a few days the suspense wore off and we forgot all about Jack. The police forgot too, for no one notified me until the news was three weeks old that Jack had been recaptured. The Doctor remained upset by recent events and was not to be pacified. His last few visits to the island had been fraught and prickly with difficulties. Now he gave up coming at all. I rejoiced to be left in peace though it was over all too briefly.

A fortnight before Christmas, a boat inappropriately named *Silver Spray* dropped anchor in the bay. This was no slim Bermudan sloop but a shabby fishing trawler with a crew of three, whose unscrupulous motive was to help themselves to a few Christmas trees from the island fir plantation. However my presence deterred them and indeed changed their scheme altogether. After coming ashore, ostensibly for water, the Skipper eyed me up and down and asked if I would like a trip up the coast next day to a settlement where he had ice to collect for packing his haul of fish.

It meant a chance to see Bob, who was working in that area, so I agreed to go on condition that we returned before nightfall.

Early next morning I surveyed the bright sunshine and light breeze with satisfaction. Affably, the Captain showed me over the trawler: the fish in the hold, the nets, the greasy galley, cramped bunks, engine and wheelhouse. He urged me to try steering the ship and having explained the course, casually left me to it, while he went down to the galley. I overheard him bellowing at the First Mate-cum-engineer-cum-cook to lay a cloth on the saloon table, and when the third monosyllabic member of the crew emerged to take over the wheel, I was called down to breakfast.

It was cramped and stuffy below and I had small appetite for the fatty hunks of half-cooked bacon heaped on my plate. Meanwhile the Skipper became more and more garrulous, while the mate appeared to be struck dumb. Knees and feet nudged against mine under the table, but I could not decide which belonged to whom. I escaped back on deck as rapidly as I could, assuring the Captain for the unpteenth time that I was neither lonely nor demented, but perfectly happy to be living alone at Aroa. He could not make head or tail of me.

Our destination lay several miles up a well-hidden inlet, and after heading towards an apparently solid range of coastal hills, a narrow gap suddenly revealed itself as we came closer. Rounding a group of pinnacled rocks, we turned sharply into a river estuary marked out with buoys. Here, the small ruffles which had disturbed the sea were smoothed for lack of wind. The water lay clear and glassy, mirroring the virginal hills on either side, except where a central current in the mainstream quickened the surface with a snake-like ripple. We chugged steadily upriver towards a small shanty town and a long wooden quay, beside which the *Silver Spray* was finally manoeuvred.

It was still quite early in the day, and having arranged a rendezvous downriver at three o'clock, I set off to see Bob, who was working on a boat in the vicinity. Astonished to find me on the mainland, he downed tools at once, took me over the motorboat which he was building, invited me to lunch and lent me his yacht tender so that I could row across the river to a favourable place for spearing flounder.

It proved a good day, spent in and out of the water, the hours melting away in summer sunshine. An abundance of flounder lay like oval shadows on the seabed, but my spearing was too tentative and half-hearted, and each time the fish darted free before I could impale them. But I was not sorry. It was pure pleasure just to wade and loiter in the cool shallows without killing anything.

Bob joined me and the afternoon ran out with the tide. We watched for the returning trawler, but by four o'clock there was still no sign of her, so we set off in a small motorboat from the boatyard to find out what delayed the *Silver Spray*. Then, rounding the last twist in the river before the settlement, we saw her bearing down on us. We waved and hailed as she approached, but she gave no answer or sign of easing speed. The Captain stared at me as though he had never set eyes on me before and for a moment we thought that he was going to steam straight past. We waved more frantically and suddenly throttling down, he beckoned us to come alongside. Bob brought his boat into a favourable position and the Skipper yelled at me to jump. It was not an easy jump, and the possibility of falling between the two boats and being crushed by them did not escape me. So I made a gigantic leap and landed sprawling on the deck.

The Captain seemed to be in a surly temper, but it was not until his mate lurched out of the galley brandishing a carving knife that I realised both men were very drunk. The third member of the crew was missing altogether and I could not help wondering with misgivings whether he had

161

been pushed overboard, or stabbed by the cook. 'He was no damn use!' the Captain pronounced ominously when I enquired, but presently I was reassured to hear that the man had been left behind, slumped at the bar.

Meanwhile we wove such an erratic course downriver towards the sea that in some alarm I went to the wheelhouse, and offered to steer. 'Can't you make that fellow put his knife away?' I suggested, indicating the mate snarling in the background. Bellowing loudly, the Captain drove him below and I was left alone on deck.

Eventually it became very quiet on board while the men slept off the effects of a long session's thirst-quenching at the hotel near the quay. I felt thankful the weather was calm, and that there was still enough daylight left to set my course for the island when we reached the open sea. As long as there were no reefs lurking just under the water, the voyage seemed foolproof. I was enjoying being in command of the ship and that sense of elation, which always possessed me on returning home, added to my high spirits. The main island and its smaller satellites lay on the skyline, dark, romantic and alluring. I needed no compass to find my course, for Aroa acted like a magnet. The bows of the *Silver Spray* pointed directly towards the island as I fixed in line an outcrop of rock with a tree on the headland beyond. Steering the ship, my imagination ran away with me. I had not been sailing merely for a day, but for months on end, if not round the world, then at least to Australia and back. And now my home port lay ahead!

Daydreams of landfall were interrupted, much to my regret, by the reappearance of the Captain. He had sobered up enough to fix lights on the ship and indeed by the time we dropped anchor in the bay it was almost dark. He angled for an invitation ashore and thinking to please me, heaped fish and provisions into the tender before rowing me to the beach.

'You'll have nothing left to eat!' I protested, but he insisted that I took his gifts. There might be a storm, he said, and I mustn't go hungry. I suspected that he was still a little drunk.

'You'd better not come ashore,' I told him artfully. 'My dogs are very fierce with strangers.'

I hoped he was too bleary and confused in the dusk to see their gaily thrashing tails as they waited on the beach, or remember their welcome the previous day. So I was abandoned on the shingle with my loot, and when I looked out of the window next morning I saw with immense relief that the *Silver Spray* had gone.

The summer brought more visitors. The island offered a safe anchorage for ships cruising round the Pacific, crossing the Tasman sea from Australia or even circling the world. One family exploring Polynesia for a year stayed a couple of days and showed me their cabin decked with treasures: wonderful shells, carved driftwood, vivid woven fabrics and brilliant feathers from tropical birds. They were educating their two young children themselves from a plentiful supply of books and appeared very happy with their nomadic way of life.

Once a naval ship dropped anchor and the Captain was piped ashore by uniformed crew. They must have been astonished to see the only resident was a barefoot girl in shorts and shirt. After drinking a cup of tea with me in the house, my caller was duly piped back aboard by his amused minions.

The publicity of the attack brought newspaper reporters ferried over by Maoris, while Bob's father, who loved visiting the island, sent me letters addressed 'Janet of The Isles, Northland' and boxes of produce from his garden. I think he hoped for wedding bells after he and Bob took me aboard *Tiara* down the coast to the Bay of Islands for an idyllic weekend but fond though I was of Bob, I could not marry

163

him. Instead we became lifelong friends and exchanged hospitality between families in later years.

But when the summer brought an occasional yacht or fishing launch into the bay for an hour or two's shelter, or a night's anchorage, I was glad to see new faces. Sometimes I was invited aboard for a meal and offered gifts of fish, fruit and newspapers, or I was taken out for a day's fishing. But it seemed a poor sport to me, to be strapped in a chair in order to torment a monster shark or marlin, and finally be photographed beside one's victim before it was thrown back in the water dead. I preferred the days when no catch was made and we sailed far out to sea, porpoises rolling in our wake and sometimes we sighted a whale spouting water, or a solitary albatross in flight and I could look back to the island rising like a golden mountain out of the water.

Half the pleasure of going anywhere was in coming home. Though I loved a chance to explore the mainland coast and days spent on the little islands amongst the wilderness of rocks and scrub and colonies of shags and mutton birds and gulls, the return was always best . . . Coming home windburnt, tired, salt and sand on the skin, fishing lines tangled, food tins empty, dogs ecstatic, to the cool quiet of the house with its familiar smells of saddlery, candles and burnt sticks mixed with the heady scent of tobacco flowers and the rich, sticky fig-tree sap drifting in from the garden. The wonder of it was that the ordinary things were extraordinary to me. I had to see and touch to be sure of Aroa's reality. I was lost to my environment and the long hot summer days moved like slow water, but at heart I knew that my days at Aroa were numbered. The Boss made it clear in his letters that I must be gone from the island before Jack was finally released from prison. I think he feared as much for his own skin as mine, for Jack had threatened him more than once.

It did not do to dwell upon the future at all, but plans had to be made. I began to think vaguely in terms of exploring the sheep and cattle country of Australia and then continuing round the world back to England, where my family and friends had been neglected far too long. If I must go, I would go in the autumn and enjoy a second summer in Europe.

When the shearing was over and the lambs were growing big and strong, I relaxed. There was time to swim again in the heated rock pools, and to collect some of the loveliest shells to take away with me: scarlet fans, long purple augers, pearly lined pauas and big white scallop shells. The long hot summer days enticed me to all my favourite places. An ocean beach with white sand-dunes kneaded by the wind, the deep cool bush-clad gullies where tame fantails fluttered and caves where surf thundered like wild bulls roaring.

At low tide, I liked the reefs best and I combed the rocks for succulent oysters or swam in the sun-warmed water, trailing home along the shore, my pockets crammed with new shells and trophies, while the dogs raced on ahead after seagulls, or danced around, begging me to throw them pieces of driftwood, or play tug-of-war with thongs of seaweed. Each day was more precious, knowing as I did that there were fewer and fewer to come.

My passage to Australia was now booked, and as word got around the district, prospective buyers came to inspect my sheep. I sold the shearing plant and wool-press. Ben was to have back his old horse, Skipper, and the Doctor had agreed to run Blacky with his cattle. Then one of the farmers, who came to look at the ewes, offered to buy both dogs, having seen how well they worked together in the pens, Glen creeping silently round the sheep while Jock bounded over their backs, barking and urging the flow forward down the race. It was heart-breaking to part with the dogs at all, but I felt that if they must go, they would certainly be

165

happier together. And so, for the last few weeks at Aroa, I lost not only my flock but my inseparable helpers.

An unearthly quiet fell over the place. No more bleating, no tame lambs coming to the garden gate, no paws scratching at the door, no thrashing tails and riotous greetings in the mornings, or reluctant bodies to be carried out of the house at night. I could not get used to the silence, and when one of the Maori fishermen asked to have the chickens and the little grey cat, the homestead seemed dead.

Perhaps the loss of so many of my animal friends made the final parting easier. The day the dogs were sold was the worst day of all. After that I felt that the final page had been turned. It was just a matter of packing and taking a last look round.

Bob had invited me to stay at his parents' home in Auckland before I sailed from New Zealand, and he offered to collect me from the island and drive me south. The evening before he arrived was still and warm. I walked all the way round the island and back along my favourite sands. There were no waves, just a gentle surge washing over the shore. The headlands at each side of the bay were reflected in the water. Seabirds flew home to roost on the rocks. The cabbage and punga trees on the hilltop stood etched against the sky. Features which I had observed so often were printed indelibly in my mind.

It was impossible to foresee what would happen to the island in the future: whether it would be commercialised by some mercenary businessman into a boatel, a fishing lodge or holiday camp; whether a rash of beach chalets would disfigure its shores, and bars, discos, gift shops and bingo halls flourish to satisfy the tourist trade. My imagination could visualise too easily the dangers of discovery. Well, if that happened, one thing was certain: I would never return. My image of Aroa was of a place wild, romantic, and incomparably beautiful. And that was the memory which I intended to keep.

The morning of my departure was tantalisingly sunny, but a fresh wind blew and the sea was choppy with white horses. However, on this occasion the element of risk in crossing to the mainland in a small open boat was one which I could face carelessly. I felt that it hardly mattered if I did not reach the other side, and I was glad that the noise of the engine and waves dashing against the bows prohibited conversation.

Beside my cases lay a sack of lobsters which Bob had caught to take back for his parents and friends in Auckland. Tactfully, he sat on the middle thwart with his back towards me, ducking his head at each onslaught of spray. I sat in the stern holding the tiller, determined not to watch the island receding behind me but concentrating on steering up and over the hummocky exhilarating waves.

I left New Zealand on a grey drizzly morning in April. Bob came to see me off, but when we had stowed my luggage aboard the ship I would not let him wait. Prolonged goodbyes were more than I could bear. And so while other passengers hung over the ship's rail waving and holding streamers to people on the quay. I crossed the deck to the empty seaward side and turned my back to the land, already trying to sever myself from the ties which tugged at me. I did not know whether I would ever return to New Zealand, and the sadness of parting could only be allayed by concentrating on the future and thinking of Australia, Europe and my English home and family. I was glad when the ship got under way.

That evening, just before dark, the liner sailed within sight of my island. I could see it lying humped in the far distance across the smooth, leaden sea, islets and reefs studding its boundary. How many times my position had been reversed, and I had stood on the cliff-tops watching toy ships creep along the horizon! But now I was being borne slowly and irrevocably away. I thought of the house, locked and deserted, the boat shut into its shed, the woolshed bolted, my horse gone from the ten-acre, the dogs and cat puzzled and restless with their new owners, and the island itself abandoned to the seabirds, cattle and the two sheep which had evaded the last muster. Right now the tide would be out, leaving the beaches patterned with ripples, pebbles polished, oysters, clams and limpets exposed on the rocks, and hanks of seaweed like bunches of dark grapes and bronzed ribbons, strewn on the sand. Such beauty and no one there to see it! I felt my heart would break with yearning. If only the island had been mine. I would have remained like a lotus eater, all will to roam gone for ever.

PART III

SMALLHOLDER

CHAPTER TWELVE

Four months after my return to England I was married in the village church of Comberton in Cambridgeshire. Jim and I had first met in Italy shortly before my departure to New Zealand. We were attending a course on Italian art and literature in Venice. Nothing would have induced me then to change my plans to emigrate but during those four years while I was abroad, our letters had shuttled to and fro across the world with welcome regularity. Our shared passion for books and writing strengthened the growing bond between us. I teased Jim for his boring nine-till-five career as a civil servant in London, only to be reminded that both Chaucer and Robert Burns had worked for the Customs and Excise. Nevertheless, to swap a barefoot lotus-eater's life on a remote Antipodean island for that of a suburban housewife in my husband's bachelor flat in Finchley seemed close to madness.

Part of me was horrified at what I was doing but the other part remained quietly determined, confident and happy. As

I waved Jim off each morning in his black coat and Homburg hat, I would giggle with disbelief at the idiocy of love. Had I not been certain that this was just a temporary phase of my life, I would have fretted like a caged animal. Instead, new dreams preoccupied me.

We had agreed to live in the country as soon as it was feasible and we both wanted to start a family. While I pictured a remote sheep-farm and six strapping sons, Jim considered our limited funds and worked out what we could afford. We possessed £1,000 between us, the furniture in the flat, books galore and Tweed, my old Border collie, who waited to be fetched from Somerset.

Those months in Finchley seem unreal. They were spent struggling with new recipes, visiting the launderette, sitting on the floor supposedly cataloguing our chaotic shelves of books, but in fact dipping into their pages and daydreaming. I would wait impatiently for the postman to bring bulging envelopes full of brochures of impossibly expensive rural retreats.

We made one bid for a semi-derelict cottage with an oast house, pond and orchard on the Kent border but other keen purchasers increased their offer beyond our limit. Farms in the commuter belt, even the smallest, were way above our reach, yet we knew that Jim would need to continue working in London to keep us solvent. I began to despair. Alone in the confined space of the flat, I longed for fresh air, green grass, trees, flowers and, of course, sheep.

I was three months pregnant when I found Miles Farm. It was March 1957, daffodil-time in the Sussex Weald. I had taken a train from London, then a bus and finally found myself walking down a long overgrown cart-track. Leaving behind traffic, houses and people, I stepped into another world as I came through a thicket of chestnut trees and into a green oasis. The farmstead was totally secluded, set in the centre of twenty-one acres of pasture land, divided into nine tiny fields bounded by woods and drifts of wild daffodils.

As I approached, I saw over the hedge on the left an orchard of apple and pear trees next to a barn, while on the grass verge ahead stood a well-top complete with bucket and chain. Behind a lilac screen on the right, sheltering in the lee of a gigantic yew tree, was a small weather-boarded cottage farmhouse with a lichen-mottled roof and a single outsize chimney. A ginger cat and an arthritic collie dog, with a wagging tail, came to greet me.

The old couple, in their seventies, who were selling the farm, welcomed me into the house, warning me to duck under the low lintel. Built in 1542, the cottage was like a doll's house. The herring-bone beams of the living-room ceiling touched my head. The inglenook fireplace took up an entire wall and by the latched door to the staircase, the brick floor was worn concave by centuries of feet. The room was congested with furniture – upright piano, grandfather clock, a settle and antimacassared chairs – but there were square-paned windows with wide oak sills looking east and west over the garden to the fields. I was shown the dairy with crocks of eggs and skimming pans of milk resting on cool slate shelves, then the kitchen with its recently black-leaded cooking range and deep horse-trough sink. Upstairs were two beamed bedrooms with tilting floorboards. Above these in the eaves, like a hawk's eyrie, was a tiny attic bedroom shaped like an ark with its miniature window looking down on a neat vegetable garden and a row of beehives.

While the farmer's wife made a pot of tea, the old man took me round the nine fields. Massive oaks and wild cherry trees dotted the hedgerows where violets and primroses bloomed on the south-facing banks. The soil was heavy wealden clay and I noticed tufts of rushes growing in damp hollows. Good fattening land, the farmer said. He had

173

milked a small dairy herd and for many years had taken the milk in churns by pony and trap to sell in the nearest town. Now he was retiring reluctantly and he pointed out the vintage tractor and a few rusty implements and tools which he could leave behind.

Back in the house, I was given a cup of tea and a generous slice of homemade fruit cake. I noticed how the interior contained a special flavour; a mixture perhaps of wood-smoke, oil-lamps and honey. The whole place enchanted me. It was humble but unique. When I walked back along the cart-track, I kept looking round at the tiny farmhouse huddled beside the great yew tree and the green island of fields encircled by a stell-like boundary of trees. I could not wait to tell Jim all about my new discovery.

On our next free weekend we visited the farm together. The wild cherry trees were in blossom and bluebells misted the woodland edges. I counted the snags. The sea was over thirty miles away and there were no hills. The land was flat and low lying, not ideal for sheep, and twenty-one acres seemed a very small patch compared with the thousand acres of Aroa. Yet Miles Farm appealed to us both. We saw it as a beginning, a nursery for a young family and our first animals.

We made an offer of £3400. When this was accepted, we were overjoyed but immediately in debt. Jim's regular salary was vital to us now.

On the 20th of June we collected the key to the farmhouse. It was not the handiest of keys to tuck into a pocket or bag, being seven inches long and as broad and heavy as the key to a church. That evening we lit the old cooking range and wept in the resulting blue pall of smoke. Then we waded round the fields through a sea of meadow grasses, ox-eye daisies, buttercups and thistles. The farmer had sold all his cattle in the spring, so now the whole farm was a hay crop waiting to be mown. We decided to start on the field nearest the barn that first June weekend.

Jim got up at 4 a.m. to tinker with our antique machinery: a 1924 Fordson tractor which needed repeated violent cranking to goad it into action, and a converted horse-drawn mowing machine with a high metal seat next to the side cutter-bar. Eagerly abandoning my efforts to paint the smoke-darkened living-room ceiling, I went out to help as soon as my chariot was pronounced ready and waiting.

I climbed into the springy iron seat gingerly and was nearly bounced off at once as the tractor bucked forward. We had gone about ten yards when the finger-bar jammed its teeth in a greedy mouthful of overlong grass. Putting the mower out of gear, we wrenched away the clogging stalks and restarted. The bumping and jolting, stopping and starting, and climbing up and down from the seat, were not the wisest activities for someone almost six months pregnant, so we changed places and I became the tractor driver while Jim struggled and swore at the ailing mower.

Doggedly we battled round that first field. Twenty-one acres now seemed like a vast prairie. The tractor belched out exhaust, drank fuel like a camel at a water-hole after crossing a desert, and roared and rattled as if every part of it was ready to shatter. Meanwhile, the mower behind vibrated with fury, chewing, choking and spitting out the grass. Instead of looking back on rows of even swathes, we saw our path marked by humps and lumps and a ragged stubble of uneven stalks – but we were winning. The unmown island shrank smaller and smaller. Rabbits began to bolt from their lonely patch of cover until at last, after twelve hours the grass was slain. Triumph was subdued by the thought of eight more fields to mow and the sight of our arms swollen with horsefly bites.

During the week Jim went to work, travelling to London each day by motorbike and train. When he returned in the evenings and had bolted a hasty meal, we resumed mowing. Even if it rained, we mowed. While I got drenched on the tractor, needing two hands to steer and change gear, Jim rode behind on the lurching mower, keeping dry under his black city umbrella.

The days were not long enough for all we wanted to do. I divided my time between painting my way round the house if it rained or while the dew was still on the ground, or working in the fields as soon as the sun emerged or a drying breeze blew. Laboriously I walked up and down the swathes of cut grass, flipping them over with a wooden-toothed rake or shaking up the wilting hay with a pitchfork. The sweet herbal scent of meadow hay and wafts of hedgerow honey-suckle pervaded the whole farm, drifting into the house through the open doors and windows.

That summer was full of new delights. Sometimes in the evenings we heard nightingales singing in the surrounding woods. We bought a hammock and hung it between a plum and a pear tree on the lawn and swung there in brief, idle moments. The garden contained an abundance of flowers: sweet williams, wallflowers, fragrant pink rambler roses and giant red-and-white hollyhocks which craned to look in through the upstairs windows. Down at the furthest end of the garden grew soft fruit, great clumps of rhubarb and runner beans, which festooned an archway of crossed sticks. There were moles, too, throwing up their earthy hummocks everywhere, and foxes which ambled across the lawn in the early mornings. After the Finchley flat, this proximity to nature was intoxicating. I radiated health and happiness as I toiled over the hay during the last months of pregnancy.

We filled the barn and loose boxes with hay, carting it in the back of Jim's old van or on a small trailer behind the tractor. When all the sheds brimmed full, we began to build a haystack. It was difficult to assess how big a base was needed and how high the rick should rise before we tapered the roof. I knew the importance of keeping up the centre and leaning the walls slightly outwards to prevent rain

176

seeping into the stack but the right symmetrical shape was not easy to achieve. Our hay-rick began to slump sideways. We tried to prop it up with poles, but as it settled and sank, the tilt became more and more alarming. It capsized one night, leaving us with no alternative next morning but to attempt to rebuild the hay into a tidier monument.

The weather had been kind to us. July and August were hot and dry but the long, calm spell ended abruptly in thunderstorms. We were woken in the night by deep rumbles of thunder and sudden brilliant snakes of lightning. At once we thought of the unprotected stack still waiting for one last load of hay to be pitchforked on to the pointed crown. Jim leapt out of bed and, like Gabriel Oak, tore outside barefoot in the teaming rain to wrestle with a flapping stack-cover. Meanwhile cowardly Bathsheba pulled the blanket over her head! Thoughts of Jim up a ladder, the gigantic chimney on the tiny house and the massive oak trees dotting the hedgerows made me certain that each flash would single out our little world for devastation.

I remembered tales of how my grandmother would cover metal objects in thunderstorms: cutlery, brass bedpans, irons, pokers. She would then open all the doors and windows to let out the devil. On the other hand, my father, a physicist, working on research at Kew Observatory, had survived unharmed in spite of developing instruments intended specifically to attract lightning. Perhaps the fact that I had been born in a thunderstorm had some curious subconscious influence. Maybe, as a farmer, I was too aware of the fatalities among sheep and cattle which storms caused. Certainly there had been one fiendish night at Aroa when the whole herd bolted in terror at the floodlit sky and a young bullock fell over a cliff to its death. So, when Jim returned drenched but unscathed, I hugged him with relief.

It took us seven weeks to make our nine fields of hay. Both the tractor and mowing machine had needed constant attention to keep them going. Blisters from forking hay had given me calloused, peasant hands. Some of the hay heated, throwing off a pungent tobacco smell which made us push our arms gingerly into the steaming interior to test the temperature, worried that the hay might ignite.

All we needed now were mouths to eat the hay, but we resolved to delay buying sheep and cattle until after the baby had arrived.

CHAPTER THIRTEEN

The birth was due on the 17th of September. A brand new Moses basket waited in a corner of our bedroom. My suitcase was packed ready for a night flight visit to hospital and, when I had a spare moment, I would go through the pile of tiny clothes, like Kanga counting Roo's vests.

My activities took little account of pregnancy. I scythed the long grass at the end of the garden, painted the attic, creosoted a portable hen ark, cleaned out a range of pigsties in which we hoped to house calves, and whitewashed the cowshed. Impatience to see animals on the farm was equalled by an accelerating impatience for the baby to be born as the third and then the fourth week of September slipped past.

The first frosts of autumn took us by surprise. In the evenings I lit a log fire and hung a big soot-blackened kettle over the flames for smoky cups of tea and coffee. The house was lit by candles and oil-lamps and we could only take a bath if we planned ahead and stoked up the old range in the kitchen which contained a back-boiler. Sometimes field mice crept under a badly fitting door to enjoy the warmth inside, while roosting birds and restless bats thumped and bumped in the roof space. Gradually I began to recognise the different sounds of the house: the rattle of loose tiles, the thrashing of roses and shrubs against the weather-boarded outer walls, the drip from a tap into the stone sink and the creak of the timber-framed building rocking in a high wind. Inevitably there were draughts which we tried to cure by hanging heavy curtains over the doors, and if the wind blew from a certain direction, great gusts of smoke billowed out from the fire to choke us. I hoped our baby would be hardy, for this was no place for the weak or fragile.

On the 4th of October I dug potatoes in the garden and that night I began to feel labour pains. The hospital was twelve miles away so we decided not to wait until morning. Our caution proved unnecessary. In spite of all my know-ledge and experience of sheep midwifery, the labour ward petrified me into a paralytic state of inaction for the next twenty-three hours. Then, at last, with the help of forceps, our 9 lb 15 oz daughter was born. A fine strong baby whose female sex was no surprise. I had guessed that it was fatal to set my heart on a son!

We called our daughter Rachel, which is Hebrew for a ewe lamb. The day Jim fetched us home was bright and sunny. I noticed how the woods had changed colour during my two weeks away. The leaves were now glowing gold. In our orchard the apples were ready to pick. There were big green Bramleys, crimson Worcester Pearmains, and rough-skinned Russets.

A week later we drove to a farm-dispersal sale with Rachel sleeping in her basket beside us like a contented dormouse. We bought six pullets and a three-year-old pedi-gree Guernsey cow for 45 guineas. She was the prettiest cow to be auctioned; a glossy golden-and-white creature with a straight back, neat udder, big eyes and a kindly feminine head. She had some long ridiculous name but I called her Tess.

She was delivered to the farm next day and we chained her in a stall in the newly decorated cowshed and gave her a liberal armful of the hay which had cost us sweat and tears and blisters. Then I took my brand-new bucket and three-legged stool and tried to milk our new purchase. Although she didn't kick, I could tell she was tense, whipping me round the head with her tail and holding back her milk. After half an hour I had only coaxed a pint from her tight udder. Despairingly I gave up, certain that my pail should have been at least half-full.

After feeding Rachel early next morning, I went out to the cowshed to try my luck again. This time Tess was relaxed and 2½ gallons of milk frothed into the bucket. She even turned her head to nuzzle and lick my arm as I sat beside her. We had made friends. Now we had milk galore, so I strained it into jugs and two big shallow pans in the dairy. When the cream had risen to the surface, I used a perforated saucer-shaped skimmer to scoop off the top layer for butter making.

Although we did not possess an end-over-end wooden butter churn, I had kept a small glass hand-churn from island days. It became an evening ritual to swop the baby and the churn from lap to lap, singing and playing with Rachel then winding and winding the churn handle. Eventually the cream began to granulate and the buttermilk separated. It was crucial not to overchurn at this stage. When the globules of butter started to coalesce, the liquid was drained off, the butter washed several times, salted to taste, then pressed and shaped between butter pats into a compact round or rectangle for the table. Among other treasures left behind in the cottage was a butter stamp which imprinted the design of a thistle when pushed down on the pat. The neat result and the rich dandelion yellow from Tess's Channel Island breeding and grassy diet made the butter look temptingly appetising.

In spite of making butter two or three times a week, we still had a surplus of milk so we decided to buy a pair of young calves to drink the excess. We went to Haywards Heath market and returned with two black Aberdeen Angus calves: a male and a female. The smaller heifer immediately recognised a bucket and drank her milk eagerly but the bull calf, whom we named Sambo or in frustrated moments, Dumbo, refused all sustenance. I tried to make him suck my finger, dipping it into the milk first, but he spat it out in disgust. I pushed his nose in the bucket while he kicked and fought. Fearful lest he should die of dehydration, I rushed out to buy a rubber calf-teat and fitted this on to an old beer bottle, but he would not accept that either. After twenty-four hours on hunger-strike, his appetite suddenly got the better of him and, as he discovered the purpose of the teat, he began to suck ravenously, nearly bunting the bottle out of my hands.

A couple of weeks later we were able to turn the calves out to grass by day. They galloped round the field with their tails held high in the air, jumping and frisking like young lambs but coming at a call in the evening for their supper. Tess had settled into a routine too, waiting for me by the gate at milking time. Once or twice, when other activities occupied me, Jim tried to take his turn at milking but found it unexpectedly difficult. Tess held back her flow while Jim pulled the teats too hard, unable to find the knack of exerting that gradual downward pressure of thumb and fingers in succession which produces a strong jet of milk. The tie of milking twice a day was added to my lengthening timetable of feeds: baby, hens (now laying three eggs each morning), calves, cow, husband. I also added a ginger kitten and a

Border collie puppy called Meg to the family. My old sheep-dog, Tweed, would join us later but was currently earning his keep on a friend's farm. All we lacked now were sheep.

We scanned the local newspapers for farm-dispersal sales, reluctant to buy stock at market where good healthy ewes might be hard to identify in a milling throng of culls and second-rate animals. When the sale of a flock near Maidstone attracted our attention, Jim took a day off and we drove to the farm in great excitement, wondering how much money we dare spend. Some 350 Romney ewes were to be auctioned, but at first the prices frightened us from bidding. They were too expensive and our only hope was to try for one of the smallest lots. Towards the end of the sale twelve tegs or two-year-old ewes, in lamb for the first time, were driven into the ring. We nudged each other. This lot was our last chance. The bidding started. I held my breath while Jim raised his catalogue to catch the auctioneer's eye. The bids crept up. Obviously we were not the only people who were desperate to buy. Suddenly the hammer fell. I wasn't even sure who had bought the sheep until I heard Jim call out our surname. The tegs had been knocked down to us for £12 a head. £144 plus £8 for a lorry to deliver them to the farm seemed like a fortune compared with prices in New Zealand. In their pen they looked fit and handsome, with long thick fleeces and woolly fringes almost hiding their eyes. As the nucleus of our future flock, I hoped they would do us well.

As soon as the twelve ewes arrived at Miles Farm, I went through them one by one. I examined their teeth and udders, pared their hooves, dosed them for worms and clipped away any dirty tail-end wool or blinding top-knots. All the tegs appeared sound and healthy, so I turned them into the field in front of the house where we could observe them easily.

At night the sheep dozed in the centre of the meadow; clustering in self-protection like the gathering of a clan in

183

some Scottish glen. We could see them from our bed in the early mornings as they woke and began to nibble at the grass. Sometimes, if a white frost rimed the pasture, the ewes carried icing on their backs and their exhaled breath steamed in the cold air.

It was now December so we put out a hay-rack and watched the sheep making trails across the frosted field to stand in rows, six-a-side, eagerly pulling out tufts of summer-dried grass. The puppy, Meg, eyed the flock with interest too, but more in the hope of a game than with any sense of drive. I struggled to teach her to sit, lie down, come to heel and walk beside me. Her concentration was spasmodic, her disposition overwhelmingly friendly and frivolous.

I trundled the pram, round the farm with me, pushing it into the cowshed while I milked, propping Rachel up so that she could see the animals. She was an easy, happy baby, full of smiles, with rosy cheeks and bright twinkling eyes. I no longer wished she was a boy. When I bathed her in a tub in front of the fire or left her kicking on a rug in the play-pen while I washed nappies in the horse-trough sink, I realised that of all the new additions to the farm, she was the peach at the centre.

CHAPTER FOURTEEN

The new year brought wild weather. Winds of over 100 mph buffeted the house. Tiles lifted from the roof, trees blew down, the hen ark was overturned, while strips of corrugated iron from the tumbledown farm buildings took flight. Indoors, we retreated at night from our big bedroom beside the massive chimney breast to the small back bedroom where the roar of the wind and the rocking of the cottage seemed less alarming.

After the gales came ice and snow. Pipes burst and we resorted to the well for drawing water, winding up the brimming bucket on a long chain. The snow brought a blessed silence and stillness, muffling the countryside and because there was now not a flicker of wind, every twig was coated with icing. The giant yew tree was so weighted down with snow on its flat branches that we found, to our horror, the calves reaching up and nibbling the poisonous needles and berries. We expected to see them stricken down within minutes and hastily drove them back into their shed. Amazingly none fell ill, but we lopped down the dangerous overhanging branches of the yew and kept the calves under cover while the weather was so harsh.

We had no room to house the sheep but they had their thick woolly jackets to keep them warm. They were hardly visible in their snowfield with white baubles clinging to their fleeces. I gave them crushed oats in a trough to supplement their diet of hay which I pulled out to the rack on a sledge. It was difficult to keep warm in the house. I had developed chilblains on my toes from the cold brick floor and chapped hands from dipping them too often in cold water. Thankfully Rachel proved hardy, bundled up in a woolly hat, coat and mitts like an Eskimo.

When the milder days returned, we set to work on hedge laying. At weekends Jim cleared the scrubby undergrowth and brambles from a length of hedge, leaving a row of uprights, mainly hazel, hawthorn, beech, oak and wild cherry. Then I took over, using a sharp billhook with a curved blade on one side and a straight one on the other. After selecting any particularly strong saplings to grow on as hedgerow trees, I began to lay the rest, chopping the uprights as close to the stool as possible, about two-thirds of the way through, until I could bend over the saplings to an almost horizontal position.

I had learnt to lay hedges in Wales where the craft is skilfully practised with stakes knocked in at regular intervals, branches pleached between these like a woven basket and a plait of hazel bonds neatly twisted along the top to hold the bent saplings in place. However, a hedge could only be as good as the materials available. There were always a few weak sections where gaps had to be filled with dead wood and newly planted cuttings, and sometimes the trees were too old and thick to be laid, or so brittle that they snapped when bent over.

My efforts at Miles Farm would not have earned much praise in Wales but at least I was able to make the hedges more stockproof while Jim cleaned out the clogged ditches and replaced broken gates. Sheep are notorious for breaking out of enclosures and we rapidly discovered the weakness of our boundaries. Where the hedges were too thin, we put up stakes and sheep netting topped with a strand of barbed wire to prevent the cattle rubbing their necks and leaning on the fences.

As soon as the two Aberdeen Angus calves were weaned from milk on to solid feed, we replaced them with a pair of Herefords. Whenever we needed more calves, it became my task to go to the local market and bid for my choice. I hated being the only female there, hearing the auctioneer bellow, 'The bid's with the lady!' whereupon all heads turned in my direction. I never knew if I had bought a bargain or been hoodwinked by the farmers and dealers around me. It was a case of learning through my own mistakes.

If a calf's navel looked raw, the animal was likely to be too young to have drunk much colostrum and was therefore prone to illness. Signs of scouring were to be avoided like the plague for, as with diarrhoea in babies, this could result in rapid loss of weight and strength. Instead, I looked for the clues to health: bright eyes, glossy coat, alert expression. The only danger of picking a lively well-grown calf – when I could afford it – was that, like our first bull calf, Sambo, it might have been suckled and bonded to its mother too long to take readily to bucket feeding. More than once we were forced to let a stubborn newcomer suckle directly from Tess. At first she kicked but with bribery in the form of oats, apples or cabbage leaves, she would tolerate strangers. The snag then was that greedy guzzlers would drink too fast and leave me short of milk for the house and since the creamiest milk comes with the last strippings, it was no use taking my share first.

On the 14th of March our first lamb was born. A big strong ram lamb. He had taken his time coming into the world, like Rachel, but I guessed that for our overfat sheep their first lambing would not be easy. I watched progress from the orchard gate, reluctant to interfere if the ewe could possibly manage unaided.

That first lamb marked the beginning of spring. I celebrated with a bonfire of hedge-trimmings and wheeled the pram round the fields picking catkins, pussy-willow, primroses, and wild daffodils to decorate the house.

During the next few weeks my high spirits slumped. Our mini-lambing was not one to boast about. Nineteen lambs were born to our twelve ewes, but five were either stillborn or died after difficult deliveries. I had never had to extract such enormous lambs from tight young tegs. One pair of twins was born late in the night after we had gone to bed and when I found them both dead in the morning for no obvious reason, I raged at myself for not having stayed up until the early hours. At that stage in the lambing, we had no spares for fostering so we scoured the district for other sheep-farmers and managed to buy a little Kerry Hill ewe lamb which the ewe accepted readily. The orphan was unmistakable because she was the only sheep in our white flock to have a black nose and two black knees.

The last sheep lambed on the 9th of April bringing our total flock number to twenty-six. When I remembered the 700 sheep I had sold on leaving the island, this tiny tally seemed like a childish game. Yet I sometimes thought that I worked harder now than I had ever worked in my life. There were three of us to care for without the luxury of household gadgets. The growing number of animals needed feeding and tending to. In winter, open fires and oil-lamps were not labour saving, while the warmer weather brought weeds, grass to mow and vegetables to plant. The routine work alone would have kept us busy, but our industry was fanned by enthusiasm and ambition to see the cottage, buildings, land and stock improved by our efforts.

There were no sheep pens on the farm so we designed some ourselves and built them with wooden rails. We constructed a gently sloping concrete floor for easy drainage, and incorporated at the centre a long swim dip. The efficient New Zealand sheep and cattle yards at Aroa had made me realise the importance of good stock-handling facilities. Slithering and sliding through mud after bolting animals was senseless, when a few weeks' work could save so much frustration. However, our small flock could not justify a woolshed or even a shearing machine.

When it was time in May to clip the sheep, we spread a big tarpaulin sheet on the ground beside the pens so that no grass seeds or dampness could spoil the wool. We bought a second set of handblades for Jim to try shearing alongside me, but he did not care for the job at all and decided he was more useful catching and turning over the heavy ewes for me.

The wool was superb: long in staple, silky to touch and full of lustre, a sure sign that our tegs were fit and fat. Their fleeces weighed over 10 lb each. I had not hand-clipped a sheep since my Cheviot days but the old method of snipping with big bold bites, keeping the blades flat against the body, gradually came back to me. The important part, as with machine-shearing, was to hold the ewe so that the skin was stretched taut, thereby reducing the risk of cutting into wrinkles.

Our woolly Romneys were not like the Scottish hill breeds of sheep. They had wool down their legs, wool round their cheeks and wool over their eyes. Then, to make the job more difficult, I could not get a really sharp edge on the blades. Sharpening shears is an art in itself. Considering there were

189

only twelve ewes to be shorn, the clipping cost me long hours of sweat and frustration.

A week later we received our first wool cheque for £20.00. This payment spurred Jim into doing the farm accounts since we had now lived at Miles Farm for a year. We discovered, that in spite of all our hard work, we had made a loss of £250!

There were so many things we needed to make life easier – a new tractor, better haymaking machinery, a ram for our flock, a washing machine, a fridge, electricity. Instead we sold three fat wether lambs for £5.00 each and bought two more calves for an equivalent sum. A few days later one of our best and biggest calves fell into a deep ditch overnight and drowned. I had always known that farming was a tough life but when the business is small and the animals almost part of the family, losses and disappointments are much harder to bear.

Tess was proving difficult to get into calf. We had had her artificially inseminated repeatedly, only to find her in season again three weeks later. We consulted the vet and he gave her various injections to improve her fertility, but all in vain. The only sensible thing left to do was to take our precious house cow to visit a real-live suitor. A local farmer with a deep-red Sussex bull assured me that his thoroughly experienced beast never made a mistake, so we ordered a lorry and sent Tess on her honeymoon. She returned looking smug. The farmer told us that if she wasn't in calf this time, it would not be his bull's fault for he had served her repeatedly. After three weeks, there were no signs of the restlessness and lowered milk yield we had learned to expect. Instead, Tess licked my arm when I milked her, in her usual placid friendly way, and we marked up on the calendar the date, nine months ahead, when her calf was due.

Meanwhile, our young sheepdog, Meg, had attracted admirers from all points of the compass. Stray mongrels and neighbours' pets queued up on the doorstep. Inevitably, when I tried to get in or out of the house, Meg shot between my legs, and sprinted for freedom. If we succeeded in capturing her for the night, we were kept awake by melancholy yowling, but if we let her go her promiscuous way, we would look out in the morning to find her knotted to the scruffiest and least desirable of her conquests.

The day Meg had her first litter of puppies she had been helping me pen sheep. In the afternoon I discovered her in the hay barn busily tunnelling out a comfortable nest for herself. That evening she produced six healthy puppies, five black and white like herself, and one looking suspiciously like a golden labrador. I resolved that it was time my Border collie, Tweed, came home. At eight years old he would make a good mature mate for flighty young Meg, and their progeny could be valuable working sheepdogs.

Tweed's return was a moment of great joy. Whether he remembered me, I could not tell, but he was as willing to work for me as ever. He had grown broader and sturdier with age, but he had lost little of his speed and none of his skill at herding sheep. The minute he saw our flock, he was down almost on his tummy, crouching, creeping and watching the ewes with hypnotic eyes.

Meg adored her new companion, flirting with him brazenly but seldom copying his exemplary work in the field or his loyalty to one person. She preferred to be Rachel's playmate, allowing herself to be pushed, pulled and tickled. She would share her puppies with any children or strangers, never growling or snapping.

As we embarked on our second haymaking, still using the old tractor and mowing machine with the converted horse cutter-bar, we began to add up what we had achieved. No financial profit for certain, yet deep satisfaction from our intimate knowledge of the nine small fields and their

191

boundary hedgerows. The tiny cottage, the land itself and the multiplying number of inhabitants made up a world which was becoming increasingly dear to us. The scent of new-mown hay, the sight of lambs playing and the song of the nightingale at dusk had lost none of their charm.

CHAPTER FIFTEEN

We did not take a holiday for the first fifteen years of our married life. When we needed so many items for the farm, to travel and stay somewhere else would have seemed an absurdly unnecessary indulgence. Besides, we could not desert the animals. As it was, Jim had too little time at home, leaving for London early in the morning and returning between seven and eight at night. His few weeks' leave were taken to coincide with lambing and haymaking while the weekends were never long enough to finish the many projects which he started.

The weather ruled our activities on the farm but it also created unexpected pleasures and diversions which made up for the lack of conventional holidays. Just a short walk away across the fields was a large and shallow reed-bound lake. During our second winter, severe frosts froze the water from one end of the lake to the other. We dug out our ancient ice-skates, unused since my childhood in Gloucestershire and Jim's undergraduate days when he had skated on the frozen flooded Backs at Cambridge. Now, unpractised, we wobbled on to the ice, laughing and clinging to each other until we found our balance. Jim had more confidence than me and gradually the old skill returned until he could skate at speed, swing round in circles and even weave backwards. I floundered and fell repeatedly, but slowly, with his help and my own determination, I mastered the leaning-forward, pushing-outward movements which enabled me to glide smoothly over the ice.

We took crusts of bread so that Rachel, muffled in rugs in her pram, could watch the hungry swans and ducks tobog-gan across the frozen lake and congregate on the bank to be fed. Sometimes those same swans flew over the farm with their long necks outstretched and their wide wings rhyth-mically beating the air.

The ice held for nearly a week and then we began to hear ominous squeaks underfoot. A crazy-paving mosaic of cracks appeared, warning us of an imminent thaw, so we put away our skates.

When we could spare a few hours in summer, my greatest pleasure was to drive to a quiet part of the coast to swim and beachcomb or to walk on the South Downs. The sea brought back so many nostalgic memories. I loved the sound of breakers, the sight of waves rolling in, rearing, curling, collapsing, and the salt seaweedy taste to the air. I would splash through the rock pools, eagerly peering at the red tentacled sea anemones and the purple-and-green hearts of sea lettuce. I cracked dry beads of bladder-wrack in my fingers and traced gulls' footprints on the sand among a scattering of shells and bootlace weed. The piping of oyster-catchers and the busy, small wader birds running and probing along the shore, brought back my life at Aroa too vividly. I was caught between happiness and sadness to regain briefly these shadows of past joy.

In May and June the flowers growing on the cliff-tops reminded me of the high spur behind my Cotswold home. Here was the same wild garden: a paradise of blue hare-bells, yellow trefoil, scabious and bedstraw. I could squander hours idling on the Downs, listening to the larks overhead, watching butterflies – Adonis Blues, Orange Tips, Small Coppers – and staring from the cliff-edge at the hypnotic sea.

A sprinkling of sheep grazed the chalk upland, small woolly-faced Southdowns, sometimes wearing bells on collars round their necks. Little groups of them congregated round the high dew-ponds where they drank and there were still a few wattle-hurdled folds in use. On one expedition I came across an old forge in a downland village where a blacksmith made Pyecombe crooks for catching sheep by the hind leg. The metal head was curved in a tight hook ending with a decorative curl. I bought one to take home but in practice I found my Scottish horn stick, for capturing sheep by the neck, much easier to use.

Perhaps my affinity for the South Downs was bred into me. On my father's side, our family came from a long line of Sussex yeomen and my odd maiden name, Scrase, was traced back to the early inhabitants of Hangleton on the Downs and to Scrase Bridge near Ditchling.

Many of Jim's forebears, including his grandfather, had been farmers too, but they came from Lincolnshire. He could remember helping at harvests and potato picking as a small child and I teased him that his eagerness to start ploughing the fields at Miles Farm must result from that arable farming background.

In the spring of 1959 we bought a brand-new, red Massey-Ferguson tractor. I was now pregnant again and after the bone-shaking Fordson on metal spade lugs, I welcomed the wonderfully smooth ride on our modern machine. Slowly we had accumulated a motley selection of implements,

195

some new, some second-hand: a two-furrow plough, a set of discs, spike and chain harrows, a flat roller, a seed distributor, a hay tedder, a hedge-cutter and a ditching machine. Jim was in his element trying out these new tools. He loved fieldwork; the smell of freshly turned earth, the chance to learn the humps and bumps and character of every square yard of our land. We took it in turns to plough and cultivate a seed-bed for the crop of kale and the new leys which we planned to grow.

The meadows were so yellow with buttercups that we realised the soil must be acid and arranged to have the whole farm limed. We carted manure from the cowshed and calf pens, spreading it directly on to the land from the trailer by throwing forkfuls in a wide arc while one of us drove the tractor along at a snail's pace. There were many jobs which required two people, but we each had our own domain. Where the animals were concerned, all decisions on management were mine, but when it came to machinery, I was glad to let Jim take over. Cranking starting-handles, changing wheels, lifting draw-bars and wrenching at stubborn nuts and bolts with spanners held no appeal for me, whether pregnant or not. However, if Jim was in London and Rachel was fast asleep or contented in her play-pen at the field edge, I enjoyed a stint of ploughing with gulls flocking in my wake, or making a striped pattern through the tilth with a set of harrows, or rolling a trim flat finish to a seed-bed.

The Wealden clay was often difficult to cultivate. Rain then drought could harden the soil into unbreakable clods and sometimes, after a wet spell, the water-logged fields became impossible to work at all. Not all our crops were successful. Thistles would shoot up amid the rye-grass and clover, and docks competed with the kale or rape. Rabbits emerged from the surrounding woods, bringing more and more of their friends and relations once they had discovered a tasty bite. Nothing we tackled was ever as easy to

196

achieve as we first imagined but we were young and fit and optimistic.

Tess had produced a handsome Sussex bull calf, the same deep-red colour as his father. Our sheep flock had multiplied to forty-two and we now owned a pedigree Romney ram called George, who had a belligerent habit of bunting us in the back of the legs when we least expected. One of my hens had gone broody so I set her on a clutch of turkey eggs, hoping to rear a batch of turkeys for Christmas. I had no inkling then of how fierce stag turkeys could become. The eight poults which hatched grew into enormous white birds, who puffed up their feathers until they looked twice their real size. At the sight of us, they would spread their proud fantails and gobble aggressively. The males turned so savage that they deliberately attacked anyone who betrayed a hint of timidity, charging forward with sharp beaks and claws outstretched. They were the first and last turkeys I ever reared. Instead, I replaced them with two black-pigtailed Chinese geese and a possessive gander who acted like a watch-dog, sounding the alarm whenever anybody called at the farm.

Not many unexpected callers ventured down our hidden lane but I was lucky to have regular deliveries from a butcher, baker and grocer. Over the years all our visitors learnt the hard way to duck under our kitchen door. When one poor man looked in to say good morning as he handed over a parcel, he raised his head too soon and fell back unconscious on the path outside!

Around the time when our second baby was due, Tess became seriously and inexplicably ill. The vet diagnosed pneumonia. We kept the cow bedded down in her shed and I tried to tempt her appetite with all her favourite food. She toyed with a few chopped apples but nothing else. I went into hospital at 4.30 one morning feeling desperately worried. In the afternoon, after a prolonged struggle, our second daughter was born. A 9 lb baby, fairer than her sister but just as perfect. We called her Sally. My disappointment at not having a boy was diminished by relief that the ordeal of labour was over and by impatience to return to Miles Farm to look after Tess. Although loss of blood made me weak and in need of a long series of iron injections, I insisted on Jim fetching me and our new baby home much sooner than the doctor advised.

I was shocked by the sight of Tess. She had become so very thin and feeble. Her lungs bubbled and rasped. She was fighting for breath. The vet recommended us to have her destroyed. It was the only humane thing to do. My mother came to help me because I was not fit to be up and about myself. I went out in my dressing gown to stroke Tess and say goodbye. The swing, which we had suspended from a rafter in the cowshed so that Rachel could keep me company while I milked, hung there forlorn. Tess's milk had long since dried up.

Back in bed I cried my heart out, listening for the dreaded rumble of the knacker's lorry and the single staccato shot. And when Rachel came into my room to ask where 'Tessy, the moo' had gone, I did not know what to say to her.

Nothing was easy that autumn. Jim had to feed and tend the stock before and after working in London while I struggled to regain my health. Sally was a bright alert little baby, but she cried lustily with three-month colic, and suffered so severely from eczema on her arms that I had to bandage them and tie mitts on her tiny hands to prevent her from scratching the red oozing skin.

It was Rachel, more than anyone, who helped me to rally. At two and a half, she was cuddly, rosy, and her hair grew in pretty tawny waves. She made me laugh, copying so many of the things I did like pretending to breast-feed her teddy, and 'milk' Meg, the sheepdog, into a calf bucket.

CHAPTER SIXTEEN

Year by year the farm expanded. We increased the original twenty-one acres to fifty by renting outlying fields. After the death of Tess, we no longer kept a house-cow but continued to rear batches of calves, on reconstituted milk, to sell as yearlings or eighteen-month-old store cattle. Each spring there were more ewes to lamb as we retained the best tegs to add to the flock. Profit from selling wool and stock was immediately reinvested in the farm in the shape of another implement, a dressing of basic slag, a load of chestnut hurdles to enlarge the sheep pens and a shearing machine to run off the back of the tractor. Mains electricity was connected to the farm and a new cooker installed in the kitchen. My mother-in-law gave me her old twin-tub washing machine, so I no longer had to wash nappies by hand in the sink.

We were busier, but I doubt if we were ever happier than in our first two years in Sussex. Neighbours on a farm beyond our boundary, who had seen me toiling in the hay fields with a pitchfork when I was seven months pregnant, began to remark that they could not remember a time when I was not either feeding a baby or expecting one. My third pregnancy ended in a miscarriage, my fourth in a beautiful daughter, Susan, my fifth in another miscarriage. When, oh when, would I ever have a boy? A farm without a son was like a field without water. If cutting a birthday cake or breaking the right half of a wishbone gave me an excuse to wish, everyone knew exactly what was in my stubborn mind. In wanting a son so badly, I was not thinking of the future as much as the present. Eventually any of the girls could farm if they chose, while a boy might just as likely decide on a totally different career. Women were as capable of being farmers as boys being nurses. The importance of the farm, while the children were young, was in the background it offered. I felt that the outdoor freedom, the company of animals, the wildlife and natural beauty could only have a benign effect. Here, a boy would thrive. Besides, we needed at least one more male in the family to help balance the bevy of females!

Meanwhile, my parents, who had become increasingly anxious at the amount of work I was heaping on myself with our expanding farm and family, decided to buy an old cottage at the end of the lane. It was a neglected place, hemmed in by trees, but there were three acres around the house, part orchard and part jungle of nettles. The last inhabitants, an elderly couple, had carried their water in buckets on a yoke from a stand-pipe.

We used our tractor and trailer to remove junk, made bonfires of everything burnable and tackled the nettles and undergrowth with the mowing machine. Jim repaired the fences and we put in a group of sheep to graze the orchard. When builders finished installing the essential plumbing, my parents moved into the cottage and at once my father set

200

to work gardening. He not only created lawns, flower-beds and a vegetable plot for himself but cycled down the lane almost daily to tend our own weedy patch. He had green fingers and loved to potter quietly from dawn till dusk, sitting on a stool amid rows of carrots and onions plucking out chickweed and docks, singling lettuces with a hoe or brooding over a smoky pyre of hedge-trimmings and leaves. He was the ideal handyman, always ready to mend things for us: broken broom handles, electric plugs, faulty toys, punctured tyres and, if we needed an extra man, he would help on the farm. My mother, small but brimming with energy, was equally invaluable, assisting with shopping, cooking and the children, while I worked in the fields or fed the stock. The family loved having grandparents nearby. A succession of little visitors found their way down the cart-track pushing doll's prams, pedalling tricycles, and later riding two or three up on Daisy, the donkey.

Daisy came to us by chance through friends of friends who were desperately seeking a home for her. She was thought to be about twenty-five years old and to have been ill-treated in her youth. Dark brown in colour with an enormous head and ears, she was gentleness itself. Children could swarm over her and under her, and ride clinging to each other, three or four in a row.

Sally doted on the donkey and learnt to ride almost before she could walk. Inevitably, low homemade jumps of fallen branches appeared in the orchard which Daisy ambled over with lazy good humour. Sometimes, when she had had enough of being ridden, she collapsed slowly on to the ground, allowing anyone on her back ample time to slide off. Then she would slump flat out, shut her eyes and with a bored sigh, feign death. The first time this happened, the children rushed in panic to find me, thinking poor Daisy was dying, but we soon realised that this was merely a pantomine to ditch her load.

We used a donkey cart which had seats on two sides and a door at the rear with steps to enter or dismount. When Rachel started to attend the village school, I sometimes drove Daisy with the other children in the cart to meet her, or I would push Susan in the pram and let Sally ride Daisy alongside me. On our arrival at the school, an excited crowd of children would cluster round offering the long-suffering donkey everything edible from toffees to chewing gum.

Back at the farm the promiscuous Meg invariably had a litter of puppies. Tweed was not favoured until 1963 when his first family of eight was born. They were enchanting look-alikes, all fluffy and black-and-white, with the same stripe down their foreheads and white tips to their tails as their parents. The children begged to keep their various favourites, but when the puppies were weaned, I hardened my heart and sold them one by one, knowing that I could not provide enough sheep work for a third dog and that to retain a Border collie in idleness was a sheer waste of its potential skill.

How I was to regret that decision! Tweed never managed fatherhood again, but died a year later in his sleep with no sign of suffering. He was fourteen, and had helped me pen sheep the previous day with all his loyal eagerness and skill. He was the best working collie I had ever had the good luck to own and train. I knew that I would miss him dreadfully, especially when I needed to bring in the sheep. Meg, for all her good nature, was disobedient, scatter-brained and incapable of concentrating on any job.

The spring when Meg and Tweed had puppies brought one of our hardest seasons. Thigh-deep snow-drifts remained heaped in field corners from Christmas until March. Hungry foxes trotted about in broad daylight and all hay had to be transported to the sheep by sledge. Then, to coincide with lambing, the torrents of spring arrived. Rain, rain and more rain. The fields were awash, the sheep disconsolate, while all the children developed coughs and colds. Sally bit the thermometer, as I tried to take her temperature, and swallowed the end. I rushed her to hospital, terrified that the glass or the mercury would kill her, only to be told that if I took her home again and gave her quarts of milk, there were unlikely to be any ill effects.

Occasionally Jim took the two older children to visit his parents in Lincolnshire and I welcomed the chance to give Susan all my attention. I was beginning to realise that the bigger the family, the less time I could give to each individual. Susan was a quiet affectionate child with beautiful dark-lashed green eyes. She was my father's special companion, for on school-day mornings, while I took the others to the village, she would stop for a second breakfast at the cottage and share his bacon. Then she would shadow him in the garden until I arrived to take her home.

The walk back from my parents' cottage was always a pleasure. According to the time of year, we could blackberry on the way or pick primroses, wild daffodils or buttonholes of honeysuckle. Behind the thick hedges, sheep munched and cattle pulled at the grass, wrapping their tongues round the stalks. The hedges must have been centuries old for they contained so many different species of trees: elder, holly, field maple, ash, oak, willow, beech, crab-apple, sloe. Round the last bend of the lane, we would see the tiny farmhouse ahead, dwarfed by the giant yew, and the well-head on the grass beyond the stable-door of the kitchen. I knew the ruts and pot-holes and corners of that cart-track so well, I could have walked home blindfold.

In 1964 my contentment was broken by an unsettling letter from New Zealand. A friend wrote to tell me that the island, Aroa, was for sale. Immediately I began to daydream and a surge of restlessness set me plotting and planning. In spite of my having asked for first refusal, should the Doctor ever wish to sell up, he had not informed me when he did so. Now the current owner, who had never lived on the island permanently, wanted to dispose of the property. I wrote for details and he named his price: £25,000 lock, stock and barrel, which included sheep, cattle, boat, land and the same humble shack of a house. I could see only one major snag: our lack of money. If we sold Miles Farm and everything we possessed I doubted if we could raise more than half the capital needed. Yet I could not let this one-in-a-million opportunity slip away.

I used all my persuasion on Jim. Surely, I argued, we could beg or borrow just enough from family and friends! It was brave and unselfish of him to agree to my wild plan. He would be the one to give up his career, to leave his roots for the unknown, to face possible bankruptcy and risk the insecurity of island life, the dangers of sea crossings, the ferrying of children and animals, the possibility of illness far from help.

We told no one of our mad scheme but made a first offer for Aroa. When that was refused, we increased the sum like reckless gamblers. This time the vendor accepted and agreed to allow us a year to sell our English farm and make arrangements to travel to New Zealand. We signed the contract, paid our deposit and waited for a copy of the vendor's signature.

There was a long ominous pause. Our solicitor informed us that the vendor had gone to sea in his yacht and his signature was therefore unobtainable. My hopes plummeted. I think I knew then that this crazy dream would never be realised. We were undecided whether to put Miles Farm on the market, uncertain now of everything. In fact, we did nothing and a couple of months later our deposit was returned. We learnt that the island had been sold for a higher sum to two Americans – who were never to set foot on the shore!

I was bitterly disappointed, no matter how often I told myself that perhaps it was all to the good. Our children would have had to go away to boarding-school. Jim might have needed to take a job in far-off Auckland to keep us solvent. We would never see our parents, and Miles Farm with all our favourite animals would have had to be sold.

It seemed as though nothing was going right now. The children had been very low with measles and then scarlet fever. I had suffered another miscarriage and despaired of ever having a son. Then Meg, who had produced yet another litter of puppies, became suddenly ill and within a couple of days was dead. Her three tiny puppies were too young to survive so Jim took them to the vet to be destroyed. We were shattered by this loss. Flirtatious Meg had been part of the family, loving and lovable to everyone.

Something had to be done. Jim had his work in London, his other life, but I needed a new challenge. Miles Farm had been perfect as a nursery, but now it seemed too confined. The cottage was not big enough for five of us and the sheep were forever breaking out of the pocket-handkerchief-sized fields. We all needed more space around us and, since the collapse of my island dream, I was homesick for the hills and the sea.

When I became pregnant for the sixth time in nine years, the search began for a cheap and remote hill farm.

CHAPTER SEVENTEEN

A quarter of an hour after midnight on New Year's Day 1966, our son, Robert, was born. When the midwife held up the baby, wrapped in a towel, and said it was a boy, I didn't believe her. I wanted to see him without any covering. I was incredulous and elated. I had even enjoyed the process of giving birth because, at 8 lb, he was smaller than his sisters and in a hurry to arrive. I should have liked a home delivery, but my rhesus-negative blood made this inadvisable, so now my only objective was to return to the farm and family with my precious bundle in the frayed and tatty Moses basket.

The three girls welcomed their baby brother rapturously. Rachel, who as a small imaginative child had invented a family of her own, once asked me if I wanted a boy. When I said yes, she told me cheerfully that I could borrow one of hers if I liked. Now there were was no need for pretence. We had the real thing. However, new babies are sleepy and not very entertaining for older children and I suspect the arrival of Daisy's son gave Sally, at least, more excitement. He was an enchantingly pretty donkey: silver grey with a well-defined black cross on his shoulders. Within an hour of birth, he was frisking on his spindly legs and shaking his top-heavy head and ears. We named him Dandelion or Dandy for short and it was never a problem to remember his age because he was as old as Robert.

These two causes for celebration might have been enough to make me forget thoughts of new horizons except that by now I was addicted to reading advertisements for farms. As soon as the weekly farming journal arrived, I would comb the property column. The wild-goose chase had begun. We went to see a farm on the Welsh Border. I loved the place. It made me think of Wuthering Heights with its whitewashed slated house half-way up a mountain, sheltering below a group of stunted Scots pine trees. There were waterfalls tumbling down the rocky hillside, Welsh ponies, and sheep everywhere. Even the name appealed to me: Taly-maes, which appropriately meant 'the end of the fields', but Jim was cautious. Some 400 high, wind-swept acres, with hardly one enclosed hay meadow, would be impractical. Besides, the price was too much for us.

We switched our search to Exmoor and looked twice at a farm which was affordable and offered scope. There were 200 acres of pasture land, a dilapidated stone farmhouse and two semi-ruined cottages which might be renovated. We dithered. It seemed a bargain but, though the land was over 1000 feet high, the farmstead was set in a boggy hollow. There was no sense of height and no view. Common sense told us that it was a good buy, but it lacked the beauty I had set my heart on, so we turned it down.

Buying a farm is even more difficult than finding the right house. If we were to put our time and energy into a new enterprise, the land had to possess a special appeal, for farming requires long-term devotion to a particular place. While arable farmers might put good fertile soil at the top of

208

their priorities and dairy farmers look for lush pastures, I wanted hills and water, space and height, a varied beautiful landscape as far from roads and towns as possible. Perhaps I was dreaming of the unattainable!

Then, in early March, just before lambing started, I read of a farm for sale in the West Country: 140 acres with hill-grazing rights on the adjoining common land and a six-bed-roomed seventeenth-century farmhouse for £13,000. It sounded too good to be true. We drove down to Somerset the following weekend.

Misreading the map, we became lost in a maze of lanes, their deep banks yellow with primroses. We stopped to eat sandwiches in a clearing beside a wood. I fed the baby, then climbed a gate into a field, intending to look over the rim of the hill to try to get my bearings. Beyond the hump, I halted, stunned by the unexpected bird's-eye view of a long steep combe winding into the hills. And there, at its heart, was Holcombe, the object of our search.

The long low sandstone farmhouse and barns were tucked into the hillside amid a patchwork of meadows and hanging oakwoods. A stream glinted in the valley bottom, twisting its way down from the high heather moorland at the head of the combe. The landscape unfolded for miles with a distant sliver of sea and a river looping towards it, far away in the vale.

Finding a place which exceeds hope and expectation is like falling in love. Response is immediate and so powerful that caution and reason are tossed aside. I rushed back to the family, wanting to share my discovery, knowing from that one quick glimpse where I now longed to live.

'I've found it!' I said and pointed from the brow of the hill. 'That's it! That's it!'

Eagerly, we drove down the lane through a hamlet of thatched cottages, past a tiny church perched on the hillside, then deep into the combe to find another lane leading back upstream. The narrow road tunnelled between overhanging trees and ended by a bridge at the bottom of the rough farm-track. We bumped over the loose stones and arrived in the yard. The farmer came out to greet us and we were ushered into a big square oak-panelled hall, where a fire of logs glowed in the wide hearth and a screen was drawn up behind a long settle to keep out draughts.

After our little cottage in Sussex, this rambling farmhouse seemed enormous and I guessed it would be cold, but in the kitchen, which had once been a cider room, there was a warm stove. A magnificent wistaria grew up the front of the house and, to the rear was a cobbled courtyard with fig trees and the gargoyle of an old man spitting out water into a stone trough. The pinkish hue of the sandstone walls and farm buildings matched the rich red soil in the lower fields, but when we were driven in the farmer's truck to the higher land, over 1000 feet up, I saw that the earth turned a peaty brown: fern and heather country. The steep slopes, the outcrops of boulders and the inaccessibility of some of the fields made me realise that this was not an easy lazy farm. It was a true hill farm but hills, however hard and difficult, were what I wanted. Buzzards wheeled overhead on tooth-edged wings, wild red deer lurked in the oakwoods and hill ponies browsed on the high moorland. The quiet combe contained everything I loved; water, grass, trees, birds, animals and wilderness.

I returned to Sussex in a fever of excitement. Two weeks later our offer for Holcombe was accepted. We were overjoyed and fearful at the same time, wondering if my bold plan could possibly succeed. Our parents were appalled, sure that the weekday separation would mark the beginning of the break-up of our marriage. But such thoughts never occurred to us; I had never questioned Jim's intention to continue working in London, quite apart from the financial need for him to do so. Nor had he ever tried to make me change direction. He accepted that hill farming was my ambition and that to take on Holcombe was my free choice and responsibility.

It was agreed that I should live in Somerset with the children and that Jim would remain to run Miles Farm during the week but join us at weekends. We had saved and inherited a little money which, together with generous family loans, would pay for the new farm without having to sell the roof over Jim's head. The prospect of deserting my parents was not quite so sad if we could keep our base in Sussex. They would caretake for us at weekends and the children and I could often return to see them. Besides, we would still be coming back to shear and make hay.

Our intention was to simplify our farming by keeping only the young stock in Sussex – a group of bullocks and our yearling ewes. The ninety-strong breeding flock would go to Somerset, also the nucleus of a single suckling herd of beef cattle in the form of six recently calved Hereford heifers which we had bought in as calves two years ago. Daisy and Dandy, the donkeys, would come with us to our new home and just enough furniture to make Holcombe habitable. The farm implements would have to be shared and swopped between the two farms.

Jim was totally supportive over this venture, as he had been in trying to buy Aroa. I knew that he was putting my happiness first. At the same time I recognised that he was independent and stoical by nature and might even thrive on a measure of solitude after a day in the city. He liked a quiet routine, simple food – bread, eggs, cheese – and on winter nights would happily lose himself in a book or work on the farm accounts and correspondence. As an early riser, he was tailor-made for the unrelenting demands of before-breakfast farming. Obviously he would miss the family, but during the week he rarely saw much of the children before their bedtime. I was more anxious about the long hours of driving he would have to endure every Friday and Sunday night. As for the challenge which I was taking on myself, I could not tell how difficult it might be until I faced it.

211

PART IV

HILL FARMER

CHAPTER EIGHTEEN

Our first night at Holcombe proved disastrous. The furniture lorry failed to arrive so we were forced to bed down on the bare floorboards. We huddled together in sleeping bags while Susan cried with colicky pains and Jim and I tossed and turned, unable to sleep for worry. The house was alive with unnerving scrabblings and creaks, and on coming in to the kitchen next morning, I found a horrible king-sized rat breakfasting on some apples left there in a box.

When the driver arrived with the furniture van, he was in a bad mood. Why, he asked, did we want to live in this God-forsaken place? The low branches in the lane had scratched his lorry and broken his wing mirror, and he had taken a dozen wrong turnings, ending up in a forest where the road fizzled out completely, before finding the right way.

He unloaded our chattels at high speed. They looked forlorn in the big rooms. We shut the doors on two bedrooms and one sitting-room, having nothing to put in them at all. Susan and Sally were to share a room, while Rachel and Robert had two small bedrooms close to our own which overlooked the cobbled courtyard. Upstairs there was a long corridor like a ship's passage with eight doors painted funereal black. It was dark and spooky and I couldn't wait to paint the woodwork white. Not until I opened our bedroom window wide to get rid of the musty disused stuffiness did my spirits lift. A dipper was bobbing on the rim of the stone trough where spring water bubbled from the gargoyle. Now I was all impatience to explore outside.

We had finished haymaking at Miles Farm and in late July we were faced with a second crop in Somerset. When we walked up the fields on our first morning, we found a herd of wild ponies feasting on the long grass and a bunch of red deer bounded over a hedge into the woods, while rabbits and hares scuttled from under our feet. No wonder the hay crop was thin round the field margins! Overhead a family of buzzards soared and wheeled; their wistful cries and the bubbling of spring water became the hallmarks of Holcombe.

The combe was even more beautiful than I remembered: a secluded tranquil place, a haven for wildlife. We found marsh orchids, ragged robin and wild raspberries. In the woods there were badger sets, noisy green woodpeckers and shy pied flycatchers. High up, where the trees gave way to rough sheep pasture, skylarks were trilling and small heathland plants like tormentil, cow wheat and milkwort studded the turf with colour. On three sides of the farm, common land adjoined the boundary: 4000 acres of grass, ferns, heather and whortleberry, running in a long ridge towards the sea – Coleridge's silver Severn sea which was seldom blue, but shallow, tidal and silty, a place for wild geese, ducks and wading birds.

There was so much variety in the landscape with native oakwoods, dark fir plantations, open moorland, steep green pastures, golden cornfields in the distant vale and beyond the pale sea, the faint blue shadows of the Welsh mountains. The enticing hilltops and the wonderful sense of space would have tempted me to walk all day but for the burden of small children.

Jim stayed to see us settled at Holcombe, then returned to Sussex to organise two lorries to transport the animals. A few days later we rushed outside on hearing heavy vehicles labouring up the steep farm-track and waited impatiently to greet our friends; the donkeys, the six cows and their calves, several coops of hens and the ninety breeding ewes. I put the stock into fields overlooked by the farmhouse so that we could watch the animals become acclimatised. They were thirsty from their journey and it was good to see how quickly they discovered the stream and the trees which offered shade.

The weather was kind to us that summer. Slowly and laboriously we made hay. We had taken over a small grey Ferguson tractor and mowed the grass every weekend. On weekdays I turned the hay with a tedder brought from Sussex while the girls looked after their baby brother. After years of farming pancake-flat fields at Miles Farm, I was totally unaccustomed to the dangers of working on steep slopes. It was a wonder I did not kill myself, slipping and sliding downhill on the tractor with no safety frame or double wheels. However, a neighbouring farmer was help-ful. He owned a hay baler which he was prepared to use on our tricky land. His own farm was hilly, so he knew what risks to take. He showed me how to bowl the bales downhill to the bottom of the fields, and how to build them into tent-shaped arks which would turn the rain until they could be carted away.

The whole family came out to the hay fields, the older girls rolling bales, Robert parked in the shade of a tree and Susan playing at houses in the hollow arks which I built. We carted the bales when Jim came home, but not without many anxious moments caused by slipping loads, knocked gate-posts and more than one jack-knifed trailer.

Everywhere we looked, there was work to be done. The fences and gates were in bad shape. The house needed decorating throughout. Ragwort, which is a poisonous plant to stock, grew in yellow clumps in many of the fields and had to be hand pulled and burnt. Thistles needed chopping before they ran to seed and flew in tiny down parachutes all over the pastures. Stones, which had broken too many mower-knife sections, waited to be picked, so visitors were roped into endless stone-gathering sorties like gangs of prisoners on hard labour. An excellent way of testing their enjoyment of our company!

The school holidays were slipping away too fast and when I could spare the time, I liked to take the children out on expeditions. We climbed to the summit of the hills, nearly 1300 feet, with Robert feeling heavier and heavier in the papoose back-carrier, and the girls taking it in turns to ride on Daisy. We picked the blue-black whortleberries growing amid the heather. The indelible purple juice stained our fingers and mouths. Wurts, as they were called locally, are known elsewhere as blueberries, cloudberries, bilberries, whinberries or blaeberries and, when cooked in a pie or crumble and served with cream they make a rich and delicious pudding. One of the nearest pubs was called the Blue Ball and now I realised that this description meant the whortleberry hill.

On sunny days the children loved paddling in the trout stream and built dams to make wallowing pools where they could float a lilo. Or I would take them to a deserted part of the coast, reached by a footpath which led through cornfields and then down a steep cliff. There the children would play on the sand and in the rock pools while I searched for coiled ammonites in the dark-grey shale. Occasionally I found another fossil called the Devil's Thumb which looks like an outsize thumb with a flattened nail on the top side. We collected shells, pebbles, and fragments of crystalline rock. We salvaged useful lengths of rope, old fish boxes, oars and planks of driftwood. When it was time to climb back up the cliff-path, we were loaded down with treasure which always included a slippery hank of bronze seaweed to hang up by the kitchen door for testing the weather by its limp or crisp texture.

At the end of August the stag-hunting season started. Until then we had never given hunting a thought. It must have been almost the opening meet when I first heard the baying hounds and the hunting-horn up the combe. There were distant shouts and a line of Land-Rovers crawled along the brow of the common land. I was too busy in the garden to take much notice, but an hour or so later I heard the clip-clop of a horse coming up the farm lane. My neighbour, dressed in his scarlet hunting-coat, raised his hat and held out a bloody hunk of venison. I refused to accept it but, not wanting to offend him, I pretended we did not like meat. I did not know then that it was the custom to offer the heart of the deer to the owner of the land where it had been killed.

A few days later the Master of the Staghounds came to call. I told him that I disliked hunting and did not want the hounds on Holcombe. I suspected that this would not make me popular with many of the local people. Stag hunting was an old tradition in the area. It was the topic of conversation in every pub and market-place. The hunt met twice a week from August until April. Then I discovered that there was fox hunting on two other days and beagling after hares at weekends.

In Sussex I had never known hounds to cross our land, but Holcombe, which reached like a long arm into the hills, was a secluded habitat for deer and foxes. I began to feel like Hazel Woodus in Mary Webb's novel *Gone to Earth*. The sound of baying hounds was 'unkit and drodsome' to me. It made me shiver to think of the stag running for its life with the pack in pursuit, while the followers yelled, cracked their whips, tooted horns, revved their vehicles, raced their motorbikes and galloped their horses up and down the hills like maniacs. When the hounds were in the valley and the huntsmen encircled the combe, all my sympathy was with their quarry.

Once I found a group of followers standing in the garden watching the progress of hounds on the far side of the combe. Whatever I said in fury was met by a stunned silence, then they all scurried off down the lane like a flock of frightened sheep. Sometimes, people unaware of my views asked which way the stag or hind had gone; then I would point vaguely in the opposite direction.

There are many misconceptions about stag hunting. A harbourer goes out the evening before the hunt to look for a suitable deer. In the morning the tufters or older hounds are brought in to single out the particular animal before the whole pack is laid on to follow the scent. A terrified deer, like a sheep, instinctively heads downhill to water, sometimes to the sea, more often into a stream in one of the deep combes. The deer is not killed by the hounds but held at bay until it can be shot by a huntsman. Many deer are poached and sometimes badly injured in the process. These the hunt takes the trouble to find and destroy.

One day when I was feeding the cattle in a lower meadow, I saw the hounds come pouring down the hillside. Ahead of them ran a hind and her half-grown calf. Knowingly she made for the water and galloped downstream, trying to disguise her scent, but the calf kept to the dry bank and spoilt her tactics. I ran to the water's edge, hoping to divert the approaching hounds, but it was useless. Their noses skimmed the ground as they raced past, baying in excitement, tongues lolling, tails wagging. Later I heard an echoing shot and knew the hind was taken.

As I came to know more of the local people, I accepted that arguing about hunting was a waste of time. They were no more likely to change their attitude than I was, and they were not barbarians. There was one favoured hind on the hills which they refused to hunt. The first time I saw her in the distance, I thought she was a white goat. Instead of the usual deep fern red, her coat was white all over except for one brown patch on her rump. Because of her light colouring, she was easy to pick out among the vegetation and I often saw her browsing in her favourite haunts. Each June she calved not far from the farm boundary until she became the matriarch of a big family group.

In September we heard stags belling in the woods at dusk. They sounded like roaring bulls. The clash of antlers meant they were fighting each other for possession of a bunch of hinds. And there were deer wallows on the farm where the stags coated themselves with muddy war-paint. The sight of a stag with a fine head of antlers and a thick rough lion-like mane has never ceased to thrill me. There were groups of deer which lived so close to the farmhouse at certain times of year that I could watch them from a window, or over the top half of the kitchen stable-door. They became accustomed to my routine journeys round the farm and in winter when I drove the tractor with stock feed along a particular track, I would see deer sitting under the trees in the woods, peacefully chewing the cud and watching me without even bothering to rise. Their coats were the colour of autumn leaves or the russet bracken. They stole our grass and occasionally broke fences but I felt it was a privilege to have them living where we could watch their natural behaviour and admire their beauty.

CHAPTER NINETEEN

When the new term started at the village school nearly four miles away, Rachel and Sally set off on foot each morning down the lane to the next farm. There a minibus met them at eight o'clock and on its meandering journey would collect about a dozen children from outlying farms and cottages. If it was fine in the afternoons I liked to walk to meet the girls with Robert in the pram and Susan on the donkey.

We passed two thatched cottages on the way. Both were pinkwashed, one pale like the gills of a mushroom, the other a deeper colour, matching the rich soil. The first was a cruck cottage with a fine arched doorway. It had once been a wheelwright's forge and the old anvil still stood in the garden. At that time a young couple lived there with two little boys close in age to Susan and Robert. Through the open doorway, which faced the sun, wafted mouth-watering smells of baking – bread, ginger biscuits, chocolate cake, treacle flapjacks. The children would all linger there hopefully.

The second smaller cottage, on the opposite side of the lane, was reached by a flagstone bridge over the stream. Under a deep tea-cosy of thatch the walls were almost hidden by a wilderness of flowers: tiger lilies, giant poppies, moon daisies and tiny sweet-scented climbing roses. A schoolmistress, now retired, had lived there for half a lifetime. She loved her garden and was usually working in it industriously but sometimes, on warm afternoons, she sat in a basket chair on the verandah of her summer-house, reading. If she saw us, she would come down to talk, peering at the children over the top of the hedge trying to remember their names and apologising for her deafness. Like my father, she loved a bonfire and practically every day there would be a spiral of blue smoke rising from a pyre of leaves and weeds.

A quarter of a mile further on was the farm where our hunting neighbour lived. His yard was chaotic, with wandering chickens, hound puppies and pigs, high heaps of steaming manure, broken implements and usually two or three horses looking over their stable doors. This last fact was well recorded by six-year-old Sally when we gave her a pony-club diary. Every day she wrote in it, 'I saw a horse looking out of the stable' or 'I didn't see a horse looking out of the stable' until after about a month of these entries, it occurred to me to explain to her that she could write about other animals or even people in a pony diary!

We were fortunate with our neighbours. They were all kind and helpful and since the lane led only to Holcombe, we felt a strong sense of security, tucked away at the end of the combe.

One cottage was being re-thatched that first autumn and the children would stop to watch the thatcher at work on their way home from school, intrigued to see him climbing up and down his ladder with great sheaves of damp straw, combing it straight, beating and pegging it into position and then trimming the edges with sheep shears.

There was always something of interest for the children on their daily trek up and down the lane: a heron fishing by the stream, speckled trout hiding under the 'pooh-sticks' bridge, hazelnuts to pick and crack open, big luscious blackberries which melted on the tongue, or maybe fingers of hot shortbread to sample from our neighbour in the cruck cottage.

In October the vet came to geld Dandelion, the young donkey. The operation was necessary to prevent a case of incest, for Daisy's son was getting frisky with his mother. Donkeys are very strong for their size and nimble at kicking. In spite of local anaesthetic, Dandy did not lose his gender without scoring a few well-aimed hits. We had three of the suckler cows dehorned at the same time. The other three cows were naturally polled. After numbing the head area with an injection, a wire saw was used for cutting off the horns. It was strenuous work which filled the air with a burning smell and caused sudden jets of blood to splatter on to our faces. The wooden cattle-handling pens were inadequate and trying to hold each cow's head in a disintegrating crush was hard. Although our cows were tame, we knew that with young children around, it was safer to remove all their horns. This also helped to stop bullying between members of the herd and made dosing or any veterinary treatment easier.

The six cows were named by the family: Rhubarb, Mandy, Looby-loo, Clover, Whitefoot and Lightfoot. The last two were called after the cows in a poem, 'High Tide on the Coast of Lincolnshire' by Jean Ingelow, a sad tale of love and drowning which I had once heard read to perfection by Sybil Thorndike. The chant of the cows' names as they were called in across the marshes was haunting: 'Come up Whitefoot, come up Lightfoot!'

225

Towards the end of the year more new arrivals joined us at Holcombe. I decided to try a different breed of sheep and bought six pedigree Kerry Hill ewes from Wales. They were accompanied by a promising Border collie puppy whom we named Brock, as his black-and-white markings reminded us of a badger. I had been without a dog too long and now that the flock was grazing over such an extensive area, help with rounding up and driving was essential.

The sheep pens and dip were as old as the cattle yard and we planned to replace them as soon as possible with a New Zealand type of layout. This would include a long sheep race with a swinging gate to separate the sheep into different groups, an entrance into the adjoining barn where we could shear or pen sheep to handle in wet weather and a circular dip. A chute would make the sheep slide down into the dip and an island in the centre would enable the operator to dunk heads and release the doors to the two draining pens at the exit. Good sheep yards make work much easier, as we knew too well after a session of slithering and chasing ewes for foot paring in the pigsties, banging our heads on the low rafters and doorways while losing the wild young hurdlers over the perimeter walls.

My impatience to have a horse to ride round the farm encouraged me to offer a temporary home to a friend's Arab stallion. Unfortunately he proved to be more of a liability than an asset. He was ridden to the farm from far across Exmoor, a journey of nearly thirty miles. Probably on his way over the moor, he sniffed mares in the air, for no barrier could keep him restrained. Again and again, he broke out of our fields until he had tracked down the pony herd which lived on the heathland. There he met the wild stallion, Waterfall, a dark, intimidating, one-eyed beast with a strange empty velvety cavern on one side of his face.

Hearing snorts and squeals one morning as I walked near the common boundary, I came upon the two stallions locked in fierce battle with an audience of mares awaiting the outcome. The horses reared up on their hind legs, with their ears laid back, nostrils flared, and teeth bared ready to lunge and bite, while striking out with their fore hooves. Their viciousness was frightening. Although I carried a halter, it was hopeless to think of prising them apart.

Perhaps because my stallion was a year or two younger than the other and in fitter condition, he began to get the upper hand. Waterfall started to back away and the more he backed, the harder the other attacked. They were lathered with sweat and I could see red weals appearing. Eventually the older horse retreated behind a bush. Immediately the other plunged triumphantly towards the band of mares and circling round them, squealing with excitement, he began to hustle them away. There was no chance of catching him now. Instead I left a field gate open and a few days later, when he returned in a quieter mood with a smaller harem, I managed to halter him and lead him to a safer enclosure and the docile company of the donkeys.

Sometimes I rode the stallion but I was afraid to venture beyond the farm on to the open hills in case I met Waterfall. This fearsome, one-eyed Casanova was so alert for new conquests that he had only to catch a glimpse or a whiff of a distant horse to come thundering across the moor, ready to fight off opponents or round up any female to add to his band of wives. I decided that the sooner the stallion was returned to his owner the better and that, instead, I would look out for a docile family pony which any of us could ride.

Fanny was the answer. She was an Exmoor cross-thoroughbred mare and came from a farm near Withypool. Her thick coat was dark brown, but her nose was the pale mealy colour typical of Exmoors, and she had the strong neck, long tail and big lustrous eyes of those first wild horses. I bought her, complete with saddle and bridle, for £120 and she proved the best creature I could have chosen for the children and the farm. She had been used as a shep-herding pony on the moor and was easy to tether, sure-footed, willing to carry any load and reliably quiet.

I no longer had to climb the steep slopes, huffing and puffing with Robert in the back-carrier. Now he rode on the saddle in front of me and when he was old enough to talk his chant was 'Faster, faster!' He loved to gallop, shrieking with joy, oblivious of risk. That first love of speed seemed to shape all his future pleasures: skiing, wind-surfing, motor-cycling, paragliding. He thrived on danger.

Every autumn the hill-pony herd was rounded up by a group of commoners. These are local farmers with grazing rights on the common. They set out on horseback, and drove the ponies along the ridge to the seaward end, then down a network of lanes to a farmyard. There the mares and foals were sorted, old ones culled, a few young females retained for weaning and branding with their owner's initials, while the majority of the foals awaited sale. An auction was held a day or two before St Matthew's Fair in Bridgwater, one of the biggest sheep-and-pony fairs in the country, where any unsold foals could be re-auctioned. Over the centuries this had also been a hiring fair like Hardy's Casterbridge and it still retained a fun-fair with fortune-tellers, coconut shies, wrestling-matches, dwarves, two-headed animal monstros-ities and stalls selling toffee-apples and, if not frumenty, all kind of other strange concoctions sold amid a reek of fried onions, and horse-dung.

I had been harbouring a wish to own a few hill ponies, knowing that we had bought, with Holcombe, common grazing rights attached to the property. Though new sheep and cattle might not thrive on the bracken-infested land and would be hard to contain on the unfenced hills, foals, born and bred there, would know their territory, where to find water, where to graze and where to shelter. So, I took the children with me to the annual pony sale. We sat high up on

228

straw bales overlooking the ring. One at a time, the foals were released from a barn and encouraged to scamper round in front of their audience. Some were so lively, they tried to escape by jumping and leaping up at the bales. I was glad we were not sitting close to the ringside.

There were three stallions running on the common at that time, Waterfall at our end, and two others where the hills jutted out towards the sea. I knew that if I could buy progeny from Waterfall, the foals would tend to return to live in the vicinity of our farm. One of the commoners, who was reputed to recognise every animal in the herd and know its life history, called out the name of each foal's sire as it entered the ring. Eagerly I craned forward when Waterfall's offspring appeared, looking for distinguishing features so that if we walked on the hills, we could identify our ponies easily.

I bid for a pretty bay foal with a dished Arab head and an even white blaze. She was knocked down to me for 37 guineas. More were sold. It was difficult to know how many others waited in the barn. The second foal I chose was not quite so well grown but she was the same bright bay, again a daughter of Waterfall, and she wore a white star on her brow. I paid the same price. Both foals were totally wild and as they were driven rearing and bucking into another building, I wondered where I could put them when they were delivered to the farm.

There was an old cattle court at the bottom of the farm-track with a stone-pillared shelter, a yard and a water trough. This, I decided, was as foalproof as anywhere. I spread a bed of straw in the covered court and put fresh hay in a rack. When the foals arrived next day they were petrified and huddled in a corner, shivering with fright. They had not yet met Sally. She spent all her spare time with them over the next few weeks. Soon they were haltered and eating carrots and apples from her hand. The bigger foal, named

Honey, became Sally's pet. Instead of returning to live wild on the hill, Sally led her, coddled her and groomed her until the day when she first climbed on to her back. After that they were always together. Long before most people would think of breaking in a pony, Sally, who was small and light, would career around the hills bareback, with just a halter on Honey's head and a rope for reins.

I had to wait until the pony sale the following year to buy foals for the other children, which they were happy to turn back later to the common as breeding mares.

CHAPTER TWENTY

The marathon trek to and from Somerset each weekend must have been a trial to Jim but he never complained. We had not enough equipment to keep everything we needed at each farm and, while we had a surplus of hay in Sussex, fodder was short for the stock at Holcombe. This resulted in slow tedious journeys, towing a trailer roped down with hay or machinery.

Sometimes Jim was stopped in the dead of night by the police, who asked him to explain where he was going and to let them see what was underneath the green tarpaulin cover. More than once he set off to cross the wide gulf between the two farms by tractor, getting stiff with sitting, freezing without a cab, and causing a queue of frayed tempers behind him. The journey took him approximately eleven hours.

Jim could not bear to waste money on buying farm clothes and made a practice of wearing out anything which was no longer needed for the office or leisure. I remember him driving a tractor to Sussex wearing yellow cricket flannels, an old Cambridge College blazer, tennis shoes, and a raincoat and cap discarded by my father. Once he was flagged down by the police with his trailer-load of hay, wearing a moth-eaten dinner jacket. I would get forlorn phone calls from some village near Winchester or Stockbridge, telling me that his vehicle had broken down and he would not be home until the early hours of Saturday. In winter there was fog to crawl through or black ice to skid over or it might begin to snow after he set out, and always there was the risk of punctures and slipping hay bales.

When he returned to Miles Farm in the early hours of Monday morning, the house would be inhospitably cold and after snatching a few hours sleep, he would need to be up again to check the stock by torchlight, fill the hay-racks, then take a quick bath and breakfast before setting off on his daily journey to London. His loyalty never faltered and those early road journeys and stints of hard labour, when we were so short of money and equipment, were heroic contributions to our dual farming enterprise.

The seasonal calendar of work – lambing, calving, shearing, haymaking, dipping, stock feeding – on two sides of England needed juggling to fit in all the jigsaw pieces. If I could find friends or relations to caretake at Holcombe briefly, I liked to take the children back to Miles Farm at half-term and during part of the longer school holidays. The young ewes had to be shorn and we usually made about 1000 bales of hay in Sussex, no longer toiling over the crop by hand but using our own machinery and the help of a neighbouring contractor with a baler.

Those summer visits to our old home were always a delight. The doll's house, huddled under the giant yew tree, had not changed. I still had to remember the low lintel and head-rapping beams, while the garden surrounding the cottage remained a vital part of its charm. There were three doors into the garden on different sides of the house which enabled us to take our breakfast, lunch or tea outside when and wherever the sun shone.

My first lambing season at Holcombe turned into an endurance test. Foxes began to take so many new lambs, not just twins but singles, two or three days old, that in desperation I pitched a tent in the lambing field and slept there at night with a wooden spoon and an empty biscuit tin to beat like a drum if I was woken by anxious bleating or if I smelt the potent whiff of a fox. We had always lambed out of doors in Sussex for want of any suitable under-cover pens, but as soon as we could make room in the partly hay-filled barns in Somerset, we housed the flock at night.

Then the lambing became so much easier. Using chestnut sheep hurdles, I could separate a ewe in difficulties and run back to the house for a bucket of warm water, soap or antiseptic gel, and a cord to deliver the lamb on to a safe, dry bed of straw. Every few hours I could sit high up on the hay bales and shine a torch down on the flock to check if all was well. The chief hazard of housing lambing ewes is mismothering, especially if there are many multiple births. Often a ewe, not yet lambed, would start licking a twin belonging to another sheep and neglect her own when born. Many a time I would find two ewes fussing over three, four or even five lambs, totally confused over which was their own true offspring. Inevitably one lamb would be rejected by both sheep and only penning for several days would strengthen the weak maternal bond.

As flock numbers grew, record keeping became more important. There were old ewes which I recognised by sight, tame ones who made themselves known and others easily identifiable by a torn ear, a black spot or some individual mark, but it was no use relying on memory for the majority. Each lamb was ear tagged with its own number, tailed with a rubber ring, castrated if a male and given a coloured mark to match that on the ewe's flank so that the pair or trio could be checked in the field. Each ewe was dosed for worms, her udder examined for mastitis, deformed teats, sores or woolliness, and her hooves, which grew like over-long toenails, were trimmed. If the lambs were well fed, they were put on to fresh grass at two or three days old. Any unthrifty lambs were given supplementary bottles and kept back in a special small paddock with access to shelter. Then there were always a few triplets, orphans or rejected lambs to be reared entirely by hand. These I kept in a row of old pigsties, where they could sit under cover if it was cold or wet, but run outside if it was sunny.

A stable provided an intensive-care baby unit with two infra-red lights for warming up chilled or weakly lambs, though the worst cases were more often revived in the kitchen. There, they could be rubbed with a towel, blown with a hair dryer or tucked up in a box in the warm oven of the Aga and perhaps fed colostrum through a stomach tube. Other sheds housed foster cases or ewes suffering complications pre- or post-lambing. Prolapses, pregnancy toxaemia, Caesareans and pneumonia were the most common problems.

During the busiest six weeks of the lambing I reckoned to lose at least a stone in weight from too little sleep and too much exercise. The house grew dustier and dirtier as I trailed mud and hay into the kitchen, cluttered the draining board with lamb bottles, syringes, rubber teats and containers of milk powder. Cooking for the family became a high-speed activity: potatoes and rice puddings baked in the oven, bacon, fish or eggs fried in a moment, which needed little preparation. My children were encouraged to cook from an early age and when the girls returned from school, I was only too pleased if they felt like making a tray of buns. They shared the excitement of lambing, helping to feed the bottle lambs and to coddle the weakly ones.

The young sheepdog, Brock, was showing promise and I tried to teach him to run in a wide arc round the sheep, to drop down at a whistle and to rise and advance when given the signal. The six Kerry Hill ewes which I had purchased from the same farm were not so successful. They produced plenty of lambs but lost too much weight on our steep land. In later years, I experimented with other breeds: Cheviots for their hardiness, Texels to improve the hindquarters of the lambs and Suffolks for easy lambing and milky ewes.

Our flock produced around 160% to 165% of lambs each spring but losses including stillborn, premature, membranes over the head, accidents and pneumonia could reduce this number by 5% to 15%. There had been one outstanding lambing in Sussex when I had not lost a single lamb or ewe but now that the flock had trebled in size, it was impossible to keep quite such a close eye on the individual sheep.

Death was my blackest enemy. I hated to see animals suffer, and hardest to bear was the loss of a ewe or lamb which I had nursed or tended for a long time. The economics of rearing bottle lambs are doubtful and many farmers opt to knock weakly lambs on the head or send them to market at a few days old. I could do neither but cosseted every lamb born, however fragile its hold on life. How was I to know whether a short life was not better than none at all?

The children delighted in pet lambs, but soon grew tired of feeding them when they realised this entailed three or four bottles a day for six or more weeks. There were always a few tame lambs who got too clever. They learnt to squeeze under gates and bleat on the doorstep for their bottles. Or they would discover the garden and the delicious taste of roses. One lamb became addicted to daffodils and wandered round decapitating every plant it could find; I puzzled over what had happened to the yellow border until I caught the culprit in action.

This tendency to stray and steal caused the downfall of many pet lambs. A few hours munching in a vegetable garden or along a lush verge can result in an acute attack of bloat, which distends the stomach of the lamb until it dies abruptly of shock or asphyxia. If I could get to a blown lamb in time, I would give it a dose of olive oil, liquid paraffin, glycerine or anything oily to disperse the gas, but sometimes I was too late and I would find, to my horror, April, Finnegan, Orfy or Snowdrop blown up like a football. Miserably I would anticipate the children's tears, repress my own if I could, and count up the wasted effort of having fed perhaps 150 bottles of milk in vain.

Barring accidents and disease, the life span of a sheep is seldom left to nature. Lambs may be sold fat or as stores for further fattening, from four months old to a year. Breeding ewes are usually culled when their teeth begin to fall out at six or seven years old. Providing they can graze and digest their food, sheep are capable of living twice as long. We kept one pet ewe, a particular favourite of the family, until she died at fourteen, while a neighbour kept another tame sheep until she was seventeen. Our current mascot is a placid bottle-reared wether who appears to regard humans as friendly two-footed sheep. It would be too treacherous to send him to market. He would walk up to his own executioner to say hallo.

The ethics of sending stock for slaughter have worried me sometimes, especially when it has meant parting with family favourites. On the other hand, dying of old age is seldom painless. Animals, like humans, become decrepit, toothless and unable to digest their food properly, so that loss of weight may trigger pneumonia or some other distressing disease. So the old breeding cows and ewes must go and the excess males cannot be kept or farms would become war-zones, full of fighting bulls and rams. As long as I can make a living from keeping a flock, my purpose is to ensure the members of that flock enjoy a peaceful, contented life, however brief or long in span.

Unfortunately rams have a suicidal tendency, particularly the most expensive pedigree ones. Whenever I treated myself to an irresistibly handsome Suffolk ram, I knew I should be lucky if it survived the tupping season. If any sheep got stuck on its back, fell in a ditch, broke a leg or succumbed to one of the innumerable mystery ailments which affect these animals, it was bound to be the new ram. Our motley collection of home-bred rams were much hardier characters, acclimatised, no doubt, to our steep north-facing slopes.

When we bought Holcombe, we took little heed of the aspect, but each winter, if there was any snow at all in the area, our land would be the first to turn white and the last to turn green. If there was heavy snow, the steeply banked lanes blocked quickly for there was no room for the snow-plough to clear the drifts. Whenever the school minibus failed to reach the next farm, the children celebrated. Our sloping fields were ideal for tobogganing and skiing, and the snow was seldom too deep to ride Fanny or Honey to fetch bread or stores. We all loved exploring the countryside when the trees were weighted with white icing and so many wild creatures could be traced by their tracks: the sheeplike slots of the red deer, the claw-edged pad marks of the badger, the smaller footprints of the fox and the crooked pattern of hopping rabbits.

I would listen attentively to the forecast if I thought snow was imminent, because it took time to fetch all the stock down closer to the hay barns. Drifted snow in the yard could make it impossible to get the tractor out of the shed. Then the only way to feed the stock was to hump bales on my back or pull them on a sledge. The gates could not be opened, the water troughs froze and I was thankful for the fast-flowing stream where animals could always drink.

One winter, when the snow took me by surprise, I hurried out to drive down the ewe flock which had been grazing on the high common. On counting them, I found, to my consternation, thirty sheep missing. We searched and searched for them, by Land-Rover, on horseback and on foot. We phoned neighbours and the police but there were no signs of the strays anywhere. After a week Jim concluded that they had been rustled but I was certain they must be alive somewhere and doggedly struggled over the snowy hills day after day, gradually widening the area of my search. Although the drifts were thigh-deep in places, on the windy tops much of the heather was half-exposed and no sheep need have starved to death.

Three weeks from the day the sheep went missing, a message reached me that a strange bunch of ewes had been spotted sitting in a neglected orchard behind a local pub on the far side of the hills. We rode over to inspect them and sure enough, they were our ear-tagged sheep, looking fat and smug, sitting under the apple trees with the bitten-off stumps of sprout and cabbage plants nearby. They were reluctant to move but with Brock's help we drove them the two-mile journey home over the high ridge and down our long steep combe.

The big farmhouse could be very cold in winter. Sometimes icicles hung from the taps and the gargoyle in the courtyard looked like Father Christmas with a beard of stalactites. If the wind blew from the north or east, we moved into the west-facing square hall and huddled round the fire, fighting for the centre of the stalls and rotating ourselves like meat on a spit. At night I heaped ashes over the burning logs so that they only needed reviving with bellows in the morning. Luckily the rigours of Miles Farm had made us hardy and we were seldom ill. Sometimes I found the dishcloth frozen to the draining board in the mornings and snow blown in under the door. While the media advised people to keep their rooms at a ridiculous 68°F, a thermometer taken round the house stuck rigidly at 40°F unless held right in front of a log fire.

One winter both the tractor and the Land-Rover seized up; barrages of drifted snow hid the tools and the log pile; the steep field tracks were like toboggan runs; and hay consumption by the hungry sheep and cattle increased alarmingly. Indoors I tried to create warmth and comfort for the children, even though we were surrounded by dripping clothes, wet gumboots and pools of melted snow. On the lawn stood a giant snowman and a bosomy snowwoman, complete with realistic tufts of pubic hair gleaned by my mischievous family from a disintegrating bird's nest.

High winds were much more frightening than snow. The slate roof of the farmhouse needed repairing. Every time a gale blew, I had to drag myself up through a trapdoor in the bathroom ceiling and clamber over the ancient pegged and bark-covered roof rafters with a bundle of twine. Then I would search among the hibernating long-eared bats and pipistrelles for the damp patches and gaping holes where slates had shifted. I struggled to yank them back into position, threading my twine through the eyelets where they had once been nailed and securing this to batons and beams until, after several winters, the roof space looked like a cat's-cradle of string.

Every wild storm felled trees in the woods. At night I would listen in terror to the slates flapping and would hear the sudden gunshot cracks of splitting tree trunks and amputated branches. The ancient shallow-rooted birches went down like ninepins while the old hedgebank beeches, unlaid for centuries, dropped in swathes. Close to the house, a trio of gigantic elms creaked ominously. The children's swing hung from one of the branches but, after one vicious gale, an adjoining branch broke off and I decided the elms were too dangerous and must be felled. They were the tallest, biggest elms I had ever seen and it was sad to watch them slain. As they teetered, we held our breath, afraid the topmost branches might strike the house. Luckily only one of the trees hit the garden fence, leaving a scattering of twigs over the lawn. We sold the timber for furniture-making just in time, unaware that within a year, Dutch Elm disease would devastate trees all over Britain so that elm wood could hardly be given away. Elm has an unpleasant sickly smell when sawn and the logs burn reluctantly.

The loss of our magnificent elms spurred me into thinking about tree planting and next autumn the children helped me to plant a steep hillside plot of Christmas trees mixed with oak, chestnut, Scots pine and naturally regenerating rowan and birch. We had to erect six-feet-high deer fencing round the seedlings and dig rabbit netting into the ground, for I soon discovered that any trees planted without these precautions would be pruned to death immediately by one sort of animal or another.

An extra dry spring made me regret my choice of site when I struggled up the steep hill with containers of water for my thirsty trees. However, the seedlings survived and grew, and each December we would march up to the plantation with spades on our shoulders to dig up the pick of the Christmas trees for friends and family.

Tree planting became one of my most satisfying activities. Each year I planted a new landmark: a group of dark blue-green Scots pines on a high knoll, a triangle of larch at the head of a gully. Larches were missing from the farm and I loved to see their emerald green in spring and candle yellow in autumn. Field corners were filled with oak, ash, beech, and lengths of hedge planted with quickthorn, sloe, hazel, and two favourites – the field maple with its vivid orange-gold autumn foliage and the whispering aspen.

I chose fruit trees to screen a new barn: cherry, green-gage, plum, eight varieties of apple and two kinds of pear. I marked strong straight saplings to grow up in the hedges and experimented with both coppicing and pollarding. There were many ancient pollarded oaks on the farm, the branches having been cut repeatedly above deer height. Now they grew like bunched antlers from immense trunks, often with bright-green polypody fern sprouting from the heart and with empty hazelnut shells wedged in the bark by nuthatches.

Coppiced trees, like ash and hazel, had been cut at the stool in the past to encourage new growth for bean-sticks, hurdles, thatching-spars and walking-sticks. The old coppice grew in dense clumps, festooned with honeysuckle – the perfect habitat for dormice, who prefer to avoid the ground and live and nest in a low tree canopy where there are nuts, honeysuckle flowers and edible shoots.

The forty steep acres of semi-ancient woodland at Holcombe were designated a Site of Special Scientific Interest. Many of the trees which blew down in storms were inaccessible for sawing or removal so they remained as habitats for insects and fungi. Here, in the least disturbed part of the farm, lived a wide variety of wildlife, from wild red deer to tiny glow-worms, noisy woodpeckers and shy pied flycatchers. The vegetation was luxuriant with drifts of purple foxgloves, tall crowns of fern and trees bearded with lichen or wearing white lips of bracket fungi.

It was impossible to wander through the woods without making discoveries. Hearing the whirring chant of a grasshopper warbler for the first time, finding a black-and-sapphire jay's feather or a violet-gilled amethyst-deceiver. Sometimes we searched the woodland floor for antlers, cast in April, but usually we came upon them by chance and carried them home in triumph.

CHAPTER TWENTY-ONE

Daisy and the donkey cart were still in action in the summer of 1969. However, the narrow lanes did not encourage me to allow the children to venture far on their own. Persuasive arguments made me relent one morning and the four children with two cousins set off down the combe.

They had not been gone more than half an hour when the telephone rang. I heard my neighbour's anxious voice. 'There's been an accident,' she said. 'Can you come quickly?'

'The children?'

'No, they're all right. It's the donkey.'

I bolted down the lane to find Daisy prostrate in a pool of blood outside my neighbour's cottage, with a knot of tearful children round her and the donkey cart abandoned on the verge. A vital part of the harness had broken so that the shafts of the cart had fallen abruptly, catching Daisy's hind leg and severing an artery. She was still alive but bleeding profusely.

While I phoned for the vet, my neighbour tried to apply a tourniquet made with ripped-up nappies. Dismayed, we heard that the vet could not get to us for two hours, so we all settled by the wayside to nurse Daisy, stroking and talking to her and reassuring the worried children while we tightened and loosened the tourniquet alternately, fearing to cut off the blood supply entirely.

Cheers greeted the arrival of the vet. He skilfully stitched the open wound. An injection of antibiotics followed. Then, after a short rest, we levered Daisy to her feet and led her home slowly. The children looked sober but Daisy appeared miraculously pleased with herself, gobbling the carrots and sugar lumps which everyone offered her.

Within a couple of weeks, she was kicking up her heels like a foal instead of an elderly 35-year-old matron. Thankfully the accident caused no lasting damage, but the donkey cart and its ageing harness were consigned to retirement.

Unlike his gentle mother, Dandelion was obstinate and wicked. He bit, he kicked and he bucked as if his aim in life was to tip all riders over his head. The braver members of the family made spasmodic use of him and we offered him as a challenge to uppish visitors. Our farrier labelled him the 'Jesus Machine', because of the sharply defined cross on his back.

The donkeys had a habit of escaping by rubbing their heads on gates until the catches slipped open. Once, they found their way to the playground outside the village school nearly four miles from the farm. Though they had never been there before, they seemed to have followed the children instinctively. In later years, after Daisy's death at forty, Dandy began to go to church every Palm Sunday. I was always scared that he might kick somebody or attempt to chew a sleeve or hat but, in fact, he did nothing worse than leave his donation in the middle of the aisle. The vicar assured the congregation that it would be cherished by his roses.

A goat joined the menagerie in 1970. She was intended to be a temporary boarder but her owner seemed reluctant to reclaim her. We soon learnt why. Dotty's indiscriminate appetite cleared the garden of roses and vegetables, the yard of any dropped paper or plastic and, if the kitchen door was left ajar, she would help herself to fruit from the bowl on the table or any letters or money left carelessly on the dresser. Nothing was safe from her. We tried tethering her but she bleated so unhappily and wound herself so effectively to a standstill on her rope that we realised that keeping her like a bound hostage was sheer cruelty.

Instead, I took her to her nuptials. The Land-Rover became her taxi as I ferried her to meet a selection of stinking, randy billy-goats. But, whenever we reached the crucial part of the ceremony, Dotty had gone off the boil and laid into her ugly suitors with her large horns.

My eldest daughter had an odd fondness for the goat and longed for her to have kids, so we persisted in our efforts to get Dotty pregnant. Had I foreseen then that Heidi, Belle and all their progeny would result, I might not have cooperated so readily. At one point we owned seven goats. Our nannies never had single offspring but produced twins or triplets. The young kids, I will admit, were enchanting: daintier and more alert and playful than lambs. They would skip high in the air and climb like children on to logs, banks and walls, as if in search of the cliffs and rocky places which wild goats inhabit.

The problems of goat keeping were diverse. Poisonous bushes appeared to ring no alarm bells, lush grass caused crippling lameness and the low-slung udders tore easily. Besides, no one in our household really enjoyed drinking the goaty milk. Finding willing buyers for neutered billies and wayward females required cunning. Sell them while young and endearing was the motto. Once I had to deliver a trio to an unfamiliar address. When I arrived, I found a terraced house with only a central door and passage leading to the long back garden. By the time I had dragged three resentful goats, a bale of hay and another of straw through the carpeted interior, I sensed the new owner's dismay.

'You will remember to give them water and plenty to eat?' I pleaded.

A few days later, driving past the terrace of houses, I glimpsed three familiar friends tap-dancing on a row of garage roofs.

Our sheepdog, Brock, adored the goats. Perhaps, because they were black-and-white British Alpines, he thought they were a strange form of collie like himself. When we had reduced the goat herd to one favourite nanny, she and Brock became inseparable, trotting upstairs to sleep together in the hay loft or following behind the ponies when the children rode on the hills. As I walked round the sheep the pair came with me, and when we all gathered whortleberries, high on the common, they darted ahead, selecting the juiciest fruit with relish.

While the children were young, an assortment of pets lived on the farm. Robert and Susan kept homing pigeons which they sometimes took to school in a hamper to release and fly back with messages. There were rabbits, hamsters, white doves and a goldfish won at a coconut shy on Fair Day, which proceeded to live in healthy celibacy for the next fifteen years. We bought Khaki Campbell and white Aylesbury ducks for the old sheep-wash pond and a flighty Muscovy who eventually reached our dinner table after landing with wet feet on an electricity pole and thudding to the ground, lifeless.

The cattle were not always easy to handle and one of my worst nightmares happened after a severe frost. I drove up the combe on the tractor with a load of hay for the herd, only to be confronted by an extraordinary sight. As the cows lumbered to meet me, they lurched and reeled as if they were drunk. Some, filing along a high path, lost their footing completely and rolled over and over to the bottom of the hill. Others floundered in the soft mud near the stream, stumbling and collapsing. I prodded them and held out tempting handfuls of hay but, though they struggled, few of them could regain their feet. Instead, they sat around dazed and stranded.

I sped home to call the vet and when he arrived I took him up the combe in the transport box along the deeply rutted track. He borrowed my gloves because it was trying to snow and by the time we reached the sick herd, our hands were almost too numb to hold the flutter valves and bottles of magnesium borogluconate to be injected into the cows. None of them had moved so we hurried from one to another. The frost had triggered off an acute attack of staggers, a metabolic disorder due to a shortage of magnesium in the blood. We pushed and pummelled the cows, urging them to rise and eventually a few responded, but the rest slumped like beached whales.

The vet left me with a crate of bottles. Twice a day I had to return to inject the helpless cows and offer them hay and water. One by one they responded, even the last cow, who had sat half in the icy stream for ten days before she regained her feet. Miraculously, though all the cows were heavily in calf, not one aborted.

During the first few years we hired a Devon bull for the suckler herd. Then we alternated between polled Herefords and Charolais, but used an Aberdeen Angus on the heifers for easy calving. Whereas I could extricate most lambs myself or take a ewe with complications to the vet for a Caesarean delivery, I could not expect to pull out a big calf alone. Calves from the continental breeds grew faster so that they sold for much higher prices than many of our native cattle, but we paid for this with difficult births. I dreaded using ropes and winches and brute force. Then, if the calf died, there was the double problem of finding a quick replacement and bribing the cow to accept the newcomer.

Returning from a rare weekend away from the farm in May, I counted the cattle first thing in the morning and realised that one was missing. Though calves were due any day, they had not yet begun to arrive. The herd was living on a hilltop field with access to steep woodland for shelter. I began to search the woods and soon discovered the missing cow sitting in a hollow. When I reached her, I saw a slimy string of afterbirth hanging from her rear, but on urging her to stand, I found that she would not budge. I felt her udder. It was hot and tight with milk. Somewhere, I suspected, lay a dead calf. Then I heard a snuffle behind me and, looking round, spotted a beautiful big sandy-coloured Charolais calf sitting under a tree. It was perfectly dry and must have been born the previous day but with such difficulty that the mother's hindquarters were semi-paralysed so that she could no longer stand.

The terrain could not have been worse, the woods being far too steep for driving a tractor. I rushed down to the farm to round up help and construct a sledge out of a corrugated-iron sheet covered with the rubber floor mat from the Land-Rover. I put this on the tractor buckrake, adding a pail, empty cider bottle and calf teat, then drove to the bottom of the wood. When we reached the cow, I managed to draw off enough milk to fill the bottle and offered it to the calf. She was desperately hungry and sucked the rubber teat with gusto. Next, we all heaved and pushed the cow on to the mat and while one helper manhandled the calf, the rest of us hauled on the sledge to start the cow's slow descent through the wood. Tree roots and fallen branches impeded progress but gradually we dragged the cow, yard by yard, to the field corner where the tractor stood. Then we slid the buckrake underneath the mat, tied the animal with ropes and gingerly continued the interminable journey down to the farmyard.

Once again the vet was called to administer various injections but the cow refused to rise to her feet. She might get up, he told us dubiously, but we would need to turn her from side to side two or three times a day to prevent cramp and sores. Every weekday morning the children, in their clean school uniforms, helped to push our hefty invalid over one way, and then on to the other side when they returned home in the evening. I heaped straw underneath her, determined my patient should not suffer bedsores, but it seemed a losing battle as one week, then ten days went by and the vet said we were wasting our time. However, I was not giving up because I had detected promising signs. Instead of her appetite failing, the cow was more and more eager for the best hay and oats I could give her. Her milk yield was increasing and I was getting adept at squeezing the milk out sideways to satisfy her daughter.

Fourteen days after she went down, the cow staggered to her feet and continued to gain strength. It took a little persuasion to teach her calf to suck a real teat instead of the bottle but when the pair were finally bonded, I watched them go out to grass with great satisfaction. Excited by freedom, the calf danced round her mother with her tail held up like a flag, while even the matronly cow gave a few skips before getting down to the more important business of wrapping her tongue round the appetising clover.

CHAPTER TWENTY-TWO

Different years and seasons are identified by special events. I remember the summer of the Clouded Yellows. These butterflies, which are quite rare in England, appeared in flocks. They danced over the hay fields like a scattering of golden confetti. A magical sight. Infinitely preferable to the spring of looper caterpillars which devoured the young oak leaves like termites. Or the sad autumn of sloes.

My brother-in-law was killed in an air crash that October and while Jim went immediately to comfort his sister, I tried to distract the children with a sloe-picking expedition. We walked to a dense spinney in the heart of the combe where the bushes were heavily laden with dusky blue-black berries as big as grapes. We filled our baskets, pricking our fingers on the sharp thorns and deliberately trying not to think or remember. Since then, I have never wanted to pick sloes for wine or sloe-and-apple jelly. The berries look like tiny luscious plums, but they dry up the mouth and, to me, they are the bitterest of fruit.

Then there was the year of adders and fires. Driving home at dusk one evening, I noticed a glow in the sky over the hills and suspected a heath fire. Presently I saw, with alarm, that the bright patch of light appeared to be centred over Holcombe. Speeding along the lane, I heard the high-pitched siren of a fire-engine ahead. My imagination ran riot. The children, the animals, the house, the barns . . . all in flames! But when I came to the bridge at the bottom of the track, I found a stationary fire-engine sucking up water from the stream. The fire was not on the farm at all, but sweeping the adjoining commonland, and then I could smell the burning vegetation.

The children were outside in the yard watching the blaze on the opposite hillside and the searchlights of the firefighters' vehicles moving around the perimeter. While areas of old woody heather were deliberately swaled each year by the commoners, accidental fires in the wrong place could do much damage, burning trees, gorse, young heather and the dry dead bracken which would spread even further in the spring from underground rhizomes. This particular fire smouldered on for several days and we were perturbed to find that it had reached our boundary and burnt several fence posts.

Even more frightening was a fire at the nearest cottage when all the thatch went up like an incendiary bomb. Luckily no one was inside at the time, but to me the cottage was never the same again for the roof rafters fell in and when they were repaired, the pitch was shallower. However, the new thatch wore a charming trade mark: a straw owl perched on one corner.

The fire-engine was called to Holcombe twice. Once to help us haul a muddy bullock out of a pond in the forest and once to put out a chimney fire. It had been so cold in the house that I had stoked up the hall fire liberally. The children and I were beginning to push back our chairs and

252

shield our faces when I noticed that the beam above the fire-place was smoking. Then I felt the wall. The plaster was almost too hot to touch. Upstairs, Sally was bathing. I yelled at her to get out and dialled 999. Then I set about smothering the fire in the grate. By the time five helmeted men ran into the house, carrying their hose, I felt rather foolish since the fire appeared to be out. On closer inspection, the men decided that a beam in the wall was still alight and they began to knock out the plaster and masonry to expose the burning timber. The room became black with smuts and grime and the floor was awash, but the fire taught us a lesson. We were never so extravagant with logs again.

I replastered the gap in the wall and repainted the hall. Painting the house was a never-ending chore. When twelve rooms, two stairways and the long passage had been covered, it was time to start again. The sitting room, known as the book-room, was now furnished and lined with crowded shelves. As they grew up, each of the children had their own room. At last we had the roof repaired so I no longer had to weave cat's-cradles of binder twine under the eaves or place saucepans strategically to catch drips.

During the summer of heath fires, there was a nasty moment when we were picking wild raspberries on a hill-side above the farm. Robert was leading the way, just a yard ahead of me, along a narrow path between thickets of rasp-berry bushes and ferns, when I heard a sudden hiss like a breath drawn in through clenched teeth. An adder was coiled on the path right in front of us, its head reared in alarm, ready to strike. Lunging forward, I dragged Robert back by the arm, while the snake, with its long black zigzag stripe, slithered away into the undergrowth.

Robert was beginning to switch his enjoyment of toy tractors and go-carts to real farm machinery and motorbikes. He and his sisters were learning farming skills from an early age. They helped with lambing and calving, rolled the fleeces at shearing time and started their first tentative attempts to shear. One by one they graduated to driving the tractor, then the Land-Rover, bumbling round the fields long before they were legally allowed on a road.

Farms are dangerous places and I shudder now to think of the risks we all took: driving bulls of uncertain disposition, riding on unstable hay loads, getting showered with sheep dip and wielding sledge-hammers, pitchforks, billhooks and chain-saws. Once Sally skidded on the tractor, bounced through a hedge and ended up teetering on the brim of a lethal slope with Robert clinging to the mudguard. Carting mangolds myself one icy morning, I tipped the tractor half-way over the bank of the stream with Robert, still a baby, clasped on my lap. That no one was ever injured, beyond the odd bruise, was pure luck. When I recollect how I rode in the wild hill pony round-ups without a hard hat and sailed a dinghy on the nearby reservoir with a crew of young children and two soggy life-jackets, I wonder how we all survived.

The fact that Holcombe was isolated and far from all traffic made any strange invasions more noticeable. Sometimes we heard the shots of deer poachers after dark or lost and bedraggled walkers knocked on the door to ask the way. On one black spooky night I took two policemen and their spades up the combe in the transport box behind the tractor. While walking on the hills earlier that day, I had found a curious branch of hawthorn pushed in the ground to look like a natural tree and I saw that the turf alongside had been removed in a rectangle and then replaced. Something or someone must be buried there!

I alerted the police and accompanied them to the site. Then I waited shining a torch while they solemnly dug and dug. If this was a grave, it looked only long enough to contain a very small corpse. A hooting owl made us jump and scuffles in the undergrowth added to the sound effects. The perspiring men threw off their jackets. Their spades struck sparks from the stones. Eventually they gave up. We had drawn a blank but we laughed, weak with relief. That something had been hidden there was certain for the earth was loose under the cut turfs. Whether it had been a small body, a cache of drugs or a gun concealed by a poacher and then retrieved we would never know. Oddly, next time I rode past that spot, my pony shied for no obvious reason.

I had discovered that Holcombe Manor was listed in the Domesday records of 1086. At that time there were seventy-five acres of pasture, fifteen acres of woodland, twelve oxen, twelve goats, four sheep, three pigs, land for two ploughs and a mill. Looking at an old tithe map, I noticed a pond marked where the stream now widened out into a silty swampy jungle beside the ruins of an old dam wall. The rotten wooden spokes of a water-wheel and the heaped stones of what had once been a mill cottage remained. Immediately I was struck by the idea of restoring the pond. The farm lacked any expanse of water and the dippers, which used to bob on the stone trough in the courtyard, had long since disappeared. Perhaps I could tempt them back, and kingfishers too.

The family approved my plan and my son, with a friend, did all the donkey-work. The stream had to be diverted, undergrowth cleared, the site excavated with a digger and the soil transported to fill in dips and hollows on other parts of the farm. The dam wall was repaired and a spillway built with railway sleepers slotted into iron girders. The sleepers were removable in case we ever needed to drain and clean out the pond. In the centre, we left an island with a coppiced alder for bird cover.

Excitement mounted when we released the stream, and water began to flood the excavated site. We knew that it would take all night to reach the top of the dam so next morning we hurried down the hill to see what had happened.

The expanse of pink muddy water, surrounded by raw red banks, was not a thing of great beauty. The island was too straight sided, so I christened it Steepholm, but at least the dam had worked, holding back this new sheet of water. A cascade poured over the top sleeper in a six-foot-high water-fall, bubbling and frothing into the stream bed below.

Nature, given the chance, takes over the environment at astonishing speed. Within a few days the mud settled and the pond water looked bright and clear. Quickly vegetation recolonised the naked banks. Kingcups, brooklime, ragged robin and meadowsweet established themselves. I planted irises, willows, marsh forget-me-nots and put a bat box in a nearby clump of trees. I watched, with pleasure, as swallows swooped and skimmed after flies on warm evenings. Small native brown trout, which had come from the upper reaches of the brook, leapt in quicksilver arcs and vanished below, spreading circular ripples.

The following year we saw both dippers and kingfishers perched on the dam while a mallard duck nested on the island and delighted us with her brood of twelve bobbing ducklings. Later, tufted duck came and a Canada goose, who worked off her sexual frustration tweaking out plants and reeds. Perhaps she hankered to build a nest but after a fortnight of vandalism, she flew off, leaving the pond littered with floating debris. A family of grey wagtails emerged, tadpoles grew into frogs and handsome blue-and-golden ringed hawker dragonflies darted among the reeds. The old sailing dinghy was repainted and used as a rowing boat to reach the island and, on hot summer days, the hardiest of us plunged into the numbing water from the dam.

My interest in natural history had begun in childhood when I roamed the Cotswold hills watching birds and animals, and collecting wild flowers to identify and press in a notebook. Now I began to record a species list for the farm which was to include more than 80 different birds, over 200 flora and 26 butterflies. Each year brought new discoveries: moschatel and tutsan in the woods, pink centaury in the old pastures, a raven's nest in a high beech tree on the 1000-feet contour, hares boxing and a dormouse snoozing in a golden furry ball in the first box I put up in our old hazel coppice.

If I walked on the hills in June, an hour after sunset, I knew a place where I could watch nightjars twisting after moths on silent wings or listen to their strange churring drum-beat. Sometimes when I returned at dusk, glow-worms starred the pathside and if I waited near a badger set, I might see the badgers emerge, scratching and stretching like humans just out of bed, before making their way down to the stream to drink. Heaps of soiled bracken were turfed out of the main entrance and new mattresses of fern brought home. Litters of cubs often trespassed in our garden, rooting, playing and squealing like pigs on the lawn at night.

Every other Wednesday for a year and a half I helped with a red-deer survey, taking the same high route round the head of the combe, then coming steeply down to my favourite meadow, ringed by a stell-like amphitheatre of hills. The young sprightly brook tumbled from a damp flush where deer wallowed in a pool edged with orchids, cotton-grass and bog pimpernel. Then it threaded along the valley floor, rippling over caddis-encrusted stones, the water dappled with light under overhanging ash trees. This sheltered cup in the hills had once been home to a drover, for the ruins of his old cottage remained, half-hidden under ivy, moss and green ferns. Buzzards frequented the adjoining woods, often nesting there, and all day long they sailed and wheeled and mewed above the meadow, watching for rabbits, mice and voles.

I carried binoculars to study the deer, endeavouring to record their ages, sex and patterns of behaviour. Sometimes I counted the points of antlered heads silhouetted against the evening sky or saw small groups browsing, unaware of my presence. If I was lucky, I might find a young deer calf, curled up in the bracken. Great startled eyes would stare up at me, but the calf would sit tight, as if hoping that it had not been noticed. Similarly, our newly born calves would conceal themselves instinctively in the hedge bottoms and in clumps of nettles and, if found, would freeze rather than run away.

Though not an early riser by nature, I often had to go out when mist still cloaked the valley, while higher up the hill-tops basked in bright sunshine under a blue canopy. All the humps in the landscape stood out like islands in a white billowing sea of cloud. Slowly, as the sun rose higher, the mist would begin to waft upwards in strange wraiths and plumes, veiling the sun so that it peered out as a pale-yellow disc and the shadowy outlines of unsuspecting deer or hill ponies would wander past me. It was worth losing sleep to find this other world: more beautiful and mysterious than when seen in the ordinary light of day.

Over the years I led a number of guided farm walks and welcomed people with a genuine interest in natural history. I also launched three surveys for a local conservation group to study woodland, streams and the density of bracken on the hills. Too many habitats were being lost. Woods were being felled, and dippers and other species were in decline. The vast area of bracken, poisonous to man and beast, which covered nearly 60 per cent of the common, needed to be controlled.

On the farm we used a minimum of fertiliser, just enough to perk up the grass for the lambs in spring and improve the hay crop. Jim, who had not known a daisy from a dandelion or a sparrow from a wren when he married me, began to share my enthusiasm for conservation. When a wood border-ing our farm was clear felled, then abandoned in neglect instead of being properly replanted and managed, he organ-ised a group of sympathetic local people to purchase the site.

We all contributed money to have the wood re-established, planting oak, ash, rowan and wild cherry in tree guards so that eventually a hilltop landmark would be restored.

During the early 1980s the land on the south-facing side of the combe came up for sale. We wanted it badly but knew that we could not afford another hundred acres without selling Miles Farm. It was a difficult decision to make. Jim had lived on his own in Sussex from Mondays to Fridays for over fifteen years. He loved the cottage and had worked hard to improve it and the land. We still kept our young stock there and made a vital part of our hay needs from the small fertile fields. However, with an option for early retirement in two or three years' time, Jim decided that the opportunity to extend Holcombe was too good to miss.

Sadly we put Miles Farm on the market and negotiated for that coveted southern slope. Jim bought himself a 10 ft × 6 ft garden shed, tucked it into a corner of my parents' orchard and lived in his tiny rural retreat four nights a week, still commuting to London each day and making that long marathon trek every weekend across England to Holcombe.

The new land offered fresh scope. We increased the suckler herd to 35 and the breeding flock to 300 with another 100 followers and rams. Grass on the south side of the combe grew a fortnight earlier than our northern slopes, so twin lambs grazed there while the singles had to take their chance on the colder hill. The new fields were less steep and the warmer aspect made them more suitable for haymaking and big-bale silage. This method of conserving winter fodder had become popular in the wetter, hilly parts of Britain. Grass could be mown one day, wilted the next, then baled into giant rolls and wrapped tightly with plastic sheeting to seal the juices and preserve the grass in a palatable nutritious form. Bayoneted on a tractor spike and unloaded on a slope, a big bale would unroll like a long carpet, allowing all the cattle plenty of headroom to feed.

259

A previous owner had bulldozed out a number of hedges on our new land and these I started to restore, planting, within double fencing, lines of quickthorn mixed with ash, oak, hazel, beech, rowan and field maple. Hares were troublesome, nipping off new shoots and where I had fenced a damp hollow to plant a small copse, the deer jumped in to fray their antlers on the young saplings. Growth was discouragingly slow but there were new discoveries to enjoy: clumps of cowslips on a hedge bank, rich orange silver-washed fritillary butterflies and, in autumn, the purple-clustered fruit of spindle trees.

All our farming methods had to become more laboursaving and efficient with the increased acreage, for we could not afford to employ any extra help. The purchase of the land included a large barn, which proved invaluable for fodder storage and housing sheep at lambing time, or holding weaned calves. A strong crush enabled us to restrain the cattle while they were dosed, tagged or dehorned. We installed two new shearing stands with overhead electric motors. Robert, Jim and I were now clipping for neighbours as well as shearing our own flock: 700 or 800 sheep each summer.

Our largest field, thirty-acre Long Breach, gave easy access on to the common where we regularly grazed yearling ewes, hill ponies and cattle. Experimentally, we mowed areas of bracken and found that if this was done in early June and repeated again in July, the ferns became weakened and stunted, allowing more palatable grasses to grow and wild flowers to spread. Heathland plants like the tiny yellow tormentil, white bedstraw and blue milkwort need space and light to flourish.

The common had always been a problem to us because of stock straying down lanes and across unfenced boundaries. The hill ponies in particular were lured by lush verges and the equine attractions of what I called the red light district over the ridge. After years of local argument, the County

Council finally drew up a management plan for the common land incorporating a stock-proofing scheme. The first grids were installed and it was hoped that these and further work would turn the tide of the decreasing numbers of common animals. At last conservationists accepted that grazing was essential if the open heathland was to be retained.

I had attended Commoners and Pony Breeders meetings for over twenty years and seen the number of graziers dwindle from twenty to five. Always we were dogged by the same problems: straying, road accidents, sheep worrying, accidental fires and the ever increasing number of tourists and visitors dropping their litter, parking in the wrong places, leaving gates open and letting their dogs run out of sight. Those interminable meetings in village pubs, choking behind a smoke-screen, listening to hunting digressions and boozy slanging-matches between farmers with opposing views had made me feel a foreigner. Now with two new young wardens on the hills, I realised that I was one of the older locals.

Everywhere was being labelled: footpaths, bridleways, picnic sites, ancient monuments. Apart from living in a listed house, the farm was now capped with letters after its name: SSSI, AONB, LFA, SDA – Site of Special Scientific Interest, Area of Outstanding Natural Beauty, Less Favoured Area, Severely Disadvantaged Area!

Sometimes I feared that the wilderness I loved so much was being eroded away and that our steep-sided combe was the last quiet place in the hills. On summer weekends sponsored walkers, riders and mountain bikers swarmed to the high moorland summit, and buzzards were not the only hang-gliders in the sky. Yet remote places are needed for other species than humans and the abundance of wildlife on our farm flourished because the land still remained an undisturbed oasis.

CHAPTER TWENTY-THREE

One by one the girls left home. Rachel, the eldest, shared my interest in natural history and became a field botanist, surveying plants on the Somerset Levels and in the Lake District National Park. She then took a job seed collecting in Brazil for Kew Gardens and later worked in agriculture. Sally became a dairy-husbandry adviser on the foothills of Dartmoor, then took her ponies and her first twenty-strong flock of sheep to a smallholding in Shropshire. Both married farmers' sons. Susan, our third daughter, set her heart at an early age, perhaps in the lambing shed, on becoming a midwife and went to nurse and marry in Devon. I was left with just Robert, the youngest of the family, a hard worker with a teasing sense of humour, devoted to the farm and animals. Obstinate as his mother, he insisted on leaving school at the first possible moment and refused a college place for further study.

When he was twenty, Robert set up his own business as an

agricultural contractor, working for neighbouring farmers, shearing, ploughing, haymaking, harvesting and helping wherever help was needed. Hoping to bribe him to stay at home more often, I bought an all-terrain vehicle, a four-wheeled, wide-tyred motorbike which would speed up hills in any weather without slipping or making ruts. The ideal transport for a geriatric, my son said, but we both made good use of it. Robert built a useful light trailer to tow behind the bike with a pen for lambs at the front and a ramp for loading ewes at the back. Now we could bring in lambing cases, sick animals and new arrivals from the steepest ground.

This was the time when there was much publicity about surplus food, and farmers were being asked to diversify. There was a building on the farm in need of repair, part cottage, and part granary and stable. We decided to restore it as a holiday cottage and save it from decaying into a ruin. The old kitchen, with its big fireplace, bacon hooks and bread oven, was retained, but the stable became a bedroom and the granary loft was converted into a sitting room, bath-room and second bedroom. We pulled down a lean-to shed outside and made a walled garden, using stones picked from the high fields. For the first time at Holcombe I had a lamb-proof enclosure for planting all the flowers lambs adore: roses, lavender, clematis.

The sunny position and view from the cottage made me almost want to change houses but now I had my own eyrie. An upstairs study lined with books and island photographs and shells. Here I wrote articles for farming journals and a regular column on country topics for a local newspaper. I had had a novel short-listed in a national competition and had won a *Sunday Times* Kenneth Allsop conservation-essay award. Since childhood, books and writing had been vital ingredients of my life.

Working at my typewriter, I could look out over the sheep pens and fields to the high hills where the woods gave way

264

to heathland and scattered holly trees were trimmed into hour-glass shapes by deer and ponies. Just above the window a telephone wire crossed the yard and there fork-tailed swallows perched all summer to dive in and out of the barns where they nested, before congregating in a crowded twittering row for their autumn migration.

When Jim moved permanently to Somerset, he converted one end of a cart-shed into a farm office. There he tackled the phone calls, forms and correspondence which farms attract, as well as launching himself into work for the church, National Farmers Union and conservation trust which he had set up to replant the hilltop wood. It was also his plan to bottle our pure spring water and as a first step towards production, he collected enough stones from the fields to build a stone beehive well-head over the source. This was inscribed on the lintel as the Fountainhead of St Francis, patron saint of water, but I felt it should have been named after my own St James!

I had not missed a lambing or shearing since becoming a shepherd in my teens but now, with a son and husband able to take over my stock work at our quietest time, autumn, I saw the chance of a freedom which I had not sampled for years. My passion for islands had never abated. I set myself a goal of visiting at least one island every autumn, beginning modestly with Lundy, Steepholm, the Scillies, Welsh islands and the Hebrides, before more ambitious journeys to the Seychelles, Cook Islands and, at last, New Zealand.

Learning that Aroa was no longer farmed, I obtained permission to return to live alone in my old home for a month. I travelled there in a state of high excitement and trepidation, wondering if memory had played tricks after a twenty-seven-year absence. On reaching the hill crest over-looking that wild Northland coast, my heart hammered, when, between dusty tree ferns, I caught a first glimpse of the island across the deep evening blue of the sea. The old spell had not broken. I was ecstatic.

Maoris at the little fishing settlement gave me a warm welcome and an old friend, now in her seventies, insisted on taking me herself in her new motor-boat to the island. Dashed by spray, we bounced over the waves to Aroa, passing familiar islets and rocks. No stock had grazed the pastures for many years and the vegetation had run wild. The house had not harboured a permanent resident since I left. When I opened the door, it was like stepping back in time; the same table and the same benches stood there. I hurried to my room, the cabin by the sea, and there was the same bed and the same small cupboard. The air was fusty and damp so I flung open the window and leaned out, breathing in the smell of the sea and listening again to the washing of waves against pebbles.

At night I was woken by a weird crooning under the floorboards and discovered a family of Little Blue Penguins nesting there. After dark the parent birds shuffled up and down the steep pebbled beach, bringing home fish for their noisy youngsters. It was like living above a rowdy seafood bar but the company, waddling past the door, made me smile.

In the morning I went out to explore. I waded through the tall grass in the garden and ten-acre paddock. The woolshed and boathouse had gone, blown away perhaps in some great gale and the fir trees, which we had planted, had been felled and abandoned. The fences and stock-yard had fallen. Manuka, with its delicate white-scented flowers, was spreading everywhere and through this dense cover the native bush was struggling to return: cabbage trees, pungas, totara, pohutukawa. Giant clumps of New Zealand flax, prickly sodom's apple and the deep foot-sinking mattress of vegetation impeded progress. It was easier to walk on the salt-burnt cliff-tops where purple clematis and white reinga lilies bloomed, or on the wide rippled sands of Waiiti where oysters still encrusted the rocky outcrops.

The high-tide line was strewn with shells. I swam in the clear water, watching gannets dive after fish, and shags perch on the reef, like monks in black habits. The wistful song of the grey warbler followed me round the island. The tame impudent fantails and the flapping white underpants of ungainly pukekos or swamp hens filled me with delight.

Here was a true wilderness. I felt that I had come home. I had all I needed: food, a few books, pen, paper and this land of wild honey, flowers, birds, pristine sands and limpid sea. A month was not long enough. The days slipped away too fast.

When I left Aroa the first time, I grieved in the belief that I would never go back. Now I was able to leave in happiness, certain that I would return.

Jim met me at Heathrow airport with an armful of flowers. It was Christmas Eve and I knew that the whole family was waiting at Holcombe to welcome me. The last lap of the journey was like crossing the sea to Aroa. The sudden glimpse of the hills ahead lifted my spirits as the lane twisted into that deep, beloved combe. Even in winter the high sheep-flecked fields looked so green, and above the hanging oakwoods and dark brow of heather, buzzards wheeled. That night, awake with excitement, I listened to the bubble of spring water into the stone trough and heard owls hooting and foxes barking.

Returning to the island and now coming back to Holcombe confirmed my belief that genuine love does not change. All my loves had lasted: for a handful of special people, for hill country, sheep, natural beauty and wild sea-bound places.

Then, in 1989, news reached me of the unexpected death of my farmer friend from student days, my first passionate love. I was devastated. He was only sixty-three, five years older than me. On his card at Christmas, he had not

267

mentioned any word of illness. Though letters and cards had been my lifeline for forty years, I had not seen him since our families had grown up.

Later, on the way home from visiting one of my daughters in Shropshire, I made a detour to the small country churchyard where my friend was buried. I searched for his grave but could not find his name on any stone. An old man was clipping the path edges so I asked if he could help. Immediately he pointed to an unmarked mound, not yet greened over, on the far side of the churchyard. The raw red earth reminded me of our soil at home which stained the sheep pink and coloured the pond after rain.

Now it was my turn to give roses. I put my small bunch down at the head of the bare hillock and wished I could sit in the grass beside the grave all day, but the ground was wet from a recent shower. Instead, I stood there, looking out from the quiet sheltered stell to the wooded ridge, the cornfields and sheep pastures where my friend had farmed for so many years. For once in my life, I was bereft of dreams.

AFTERWORD

*T*he *Sheep Stell* was first published more than twenty-five years ago, so an update may be of interest to readers of this new edition.

Tragically my husband, Jim, died aged seventy-one from a sudden heart attack while cycling for charity in Cuba. We had been married for forty-two years and had planned to meet in Havana for a holiday after the ride. Instead, he came home in a metal box like a filing cabinet and I received a letter from him, written at Heathrow, offering tips on where to go for a snooze while waiting for a night flight. It was my birthday that week and he sent all his love.

The shock of this numbing loss in 1998 brought my big affectionate family rallying round and a month later I was asked to babysit at my son's cottage just down the lane while his wife went into labour at the local hospital. I stayed the night and as I was getting breakfast, I smelt smoke. I had left the vent of the wood burner open and set light to the chimney! Rushing outside I saw sparks flying over the thatch. Dialling 999 and explaining to little Tom and Lucy what had happened, they grabbed their yellow toy firemen's helmets and ran into the garden to watch for the fire-engine. Then the phone rang with news from Robert that a beautiful baby daughter had been born. I dare not tell him that I had just set his home alight and when I told the children that they had a brand new sister, they were far more excited by the approaching alarm bells ringing along the lane. The cottage was not burnt down and I thought how Jim would have laughed and celebrated the birth of Helen, our thirteenth and last grandchild. How I wished he could have seen her and lived to enjoy all the fun, interests and achievements of this new generation.

The youngsters helped me so much during the hard years of early bereavement with their singing, dancing, playing of musical instruments, sporting activities, various hobbies and obvious pleasure in staying at the farm.

Jim would have been proud too of all our grandchildren's success as adults. Apart from the youngest, who opted to travel first, all studied at university. Three girls followed in his footsteps to Cambridge, like my father who taught himself Latin and Greek to win a scholarship. Two read Medicine and one chose Law. Others studied Classics, Physics, Geography, Biology, Mathematics, Sports Science, Agriculture and Veterinary Medicine.

During gap years and long vacations they travelled all over the world. One worked as a ski instructor in Japan, two helped with outdoor activities on summer camps in the USA, one taught English in Bolivia, the Junior Doctors went to hospitals in South Africa, Romania and Malaysia, the young lawyer worked in Madrid, Brussels and Washington, the agricultural student, a keen horse rider like her mother, represented England in several international Endurance events, while others explored far and

270

wide to the tip of South America and, inevitably, New Zealand and my island.

My son, Robert, became my business partner on the farm. We had given him a plot of land after he married and his thatched cottage is often assumed to be genuinely old. The chimney and fireplace stone, the oak door and beams and the hazel spars all came from our land and woods. Surrounded by a garden with roses round the porch and walls, it blends in with the local architecture. My eldest daughter also lives on the perimeter of the farm in an old cottage set on a sunny hillside with a seven-acre wood, a paddock for her two pet Jacob sheep and stunning views of our land. Both she and her husband, a farmer's son, often help us on the farm at busy times.

Hill farming is not an easy way to make a living and it became tougher when outbreaks of foot and mouth, blue-tongue and tuberculosis restricted livestock movement. One year seven cows with TB went for slaughter and I was left with five newly born calves to bottle feed. Sheep ticks, rife in the hills, are a danger to stock and humans, causing paralysis, fever and Lyme disease. Lamb prices, like the weather, are always unpredictable. Payment for conservation work has helped us survive and though age is catching up with me, I still run guided natural history walks, check sheep daily on a quadbike with my eager young collie, Moss, keep the stock records, fill in endless Defra forms and welcome guests to our holiday cottage.

I hope Jim would applaud our progress. We have expanded the farm considerably, renting a large block of hill land, rising to over 1200 feet, adjacent to our fields and the 4000-acre common. Then, when a neighbouring farm came up for auction, I was lucky to purchase five old hay meadows on our boundary. They contain a rich mix of flora, including cowslips, yellow rattle, trefoils and orchids. Overgrown hedges of hazel, blackthorn and maple, studded with mature

271

oak, ash and wild cherry, offer a good habitat for dormice. When the pasture is grazed down in autumn, waxcaps of all hues flourish there. We also started a small flock of pedigree rare breed Whiteface Dartmoors to run on the hill, in addition to our commercial North Country Cheviot X Texel and Suffolk sheep. A herd of Hereford suckler cows and eight part-Arab hill ponies complete the stock. Farming almost the whole combe gives great pleasure. The views are a delight and each year I find new species of fauna and flora to add to my records.

Renovating the old granary for self-catering accommodation had been Jim's idea and it has proved rewarding, bringing us many interesting people, new friends and sometimes winter tenants, particularly artists and writers. The late Jenny Diski wrote much of her book *On Staying Still* here, describing finding 'the perfect cottage' and a guilty walk with my sheepdog, Ben, who led her astray in his eagerness to round up sheep. David Cairns, writing a biography of Mozart, came one Christmas and on arrival surprised me by knocking at the door and asking to borrow some clothes. Astonished, I asked what kind of clothes he needed, but he burst out laughing. My poor hearing had misheard. He wanted cloves! Another favourite guest was an eighty-four-year old moth expert who brought his moth trap and allowed Robert's children to help release his captives each morning, so now we look at Emerald Greens, Heart and Darts, Buff tips, Silver Underlings and so many other wonderfully patterned moths with different eyes.

Going on holiday ourselves is always difficult with farm stock needing daily care but I usually try to rent a cottage somewhere enticing to share with various members of the family. The Scilly Isles, the Pembrokeshire and SW coastal paths, the Hebrides and Pyrenees have been some of our destinations. Old haunts have lured me back too. Returning to the Slad valley with a couple of daughters took us to The

Woolpack Inn for lunch, a peep at my lovely old Cotswold home and a walk up Swifts Hill where I had wandered after school finding so many orchids, harebells, cowslips, white violets and clouds of butterflies. It is now a nature reserve and rightly so. A place of tranquillity where larks sing over a carpet of flowers.

One summer we went to Northumberland where it borders the Cheviot Hills. We walked up the beautiful Harthope valley. While some of the young energetic members of the family climbed to the high summit of Cheviot, I followed the burn which tumbles down from the open hills. I was back in familiar country, only a couple of miles as the crow flies from my first shepherd's cottage and hirsel. There were Blackface sheep grazing amid purple heather, curlews calling and a scattering of stone-walled sheep stells. It was like coming back full circle to the enchanted place where I first became a shepherd. Later a friend sent me a photo of my old home. It had been uninhabited for years and fallen into disrepair, used only as a trekkers' bothy, too lonely for modern taste but to me a place of great happiness.

In 2004 I decided to celebrate my seventy-fifth birthday by spending a month back on my New Zealand island. I flew to Auckland in twenty-three hours, instead of that first six-week voyage, and headed north in a much smaller plane. Here vintage friends met me, took me to a local store to stock up with food, candles, matches and essentials and ferried me to the island the following day.

Marooned, I lived in a dream, mesmerised by the beauty around me. Although the old house was now condemned as unsafe, I found my bedroom door into the garden ajar and explored inside, rejoicing to discover the Little Blue Penguins still in residence under the rotting floorboards. Outside, the corner of the bay, a warm sun pocket sheltered by a headland, became my favourite swimming place again.

273

All the familiar delights were around me: a treasure trove of shells, oysters on the rocks, giant flax bushes, shaggy cabbage trees, the haunting song of the grey warbler, the switchback path past the wild bees' nest to the ocean sand-dunes, seabirds calling, vistas of green islets and the sea in all its moods and vivid colours encircling me.

It was a magical time enhanced by visits from old friends bearing gifts of fruit, vegetables, fresh bread and milk. The son of my postman, Ben, arrived in his fishing boat with his own fifty-year-old son, to give me a big hug and chat about old times. Bob and his fellow boatbuilder, now in their late seventies, who had looked after my animals while I was in hospital and helped in so many ways, sailed out to see me. We mused over what had happened to Jack but no one knew. His life after prison remained an unsolved mystery. The widow of one of the Irishman's sons, who had returned to clip my sheep, came with some of her family and gifts of fish caught en route. All my local friends were now in their seventies or eighties but they launched their boats to visit me bringing things I might need or enjoy from frying pan and loo roll to chocolate and peaches. So much kindness and affection plus the incredible beauty around me made my return to the island unforgettable.

I am too old now to go back again but one day last year a mystery package arrived from a stranger in New Zealand. Inside was a memory stick. The contents had been found during the house clearance of a deceased relative in Queens-town. It proved to be a short cine film taken in 1954. I was astonished to see myself loading my first sheep into a boat with Ben, leading a frightened horse up the beach off the Doctor's disintegrating raft, releasing his pheasants on the island, fishing for snapper and sailing past leaping dolphins.

Looking back on exciting times is good but my recipe for happiness is live for today. This morning I saw twelve new ducklings bobbing behind their mallard mother on the pond, swallows are repairing their nests in the barns, young lambs and calves jumping and playing like children and as I passed a woodland edge I caught a sweet waft of blue-bells. Small miracles but there for everyone who cares.

Once, when I was a child, I found a four-leaf clover in a field. It is pressed still inside an old diary. I am not supersti-tious but I recognise the good luck in my life: my wonderful family, the farm and the remote glorious places where I have worked. *The Sheep Stell* is a tribute to them all.